Analysis of the mechanical response of impact loaded composite sandwich structures with focus on foam core shear failure

Vom Fachbereich Produktionstechnik

der

UNIVERSITÄT BREMEN

zur Erlangung des Grades

Doktor-Ingenieur

genehmigte

Dissertation

von

Dipl.-Ing. Tim Berend Block

Gutachter: Prof. Dr.-Ing. Axel S. Herrmann

Prof. Dr.-Ing. Hans-Günther Reimerdes, RWTH Aachen

Tag der mündlichen Prüfung: 04. August 2014

Science-Report aus dem Faserinstitut Bremen

Hrsg.: Prof. Dr.-Ing. Axel S. Herrmann

ISSN 1611-3861

Bibliografische Information der Deutschen Nationalbibliothek

Die Deutsche Nationalbibliothek verzeichnet diese Publikation in der Deutschen Nationalbibliografie; detaillierte bibliografische Daten sind im Internet über http://dnb.d-nb.de abrufbar.

ISBN 978-3-8325-3853-8

Logos Verlag Berlin GmbH
Comeniushof, Gubener Str. 47,
10243 Berlin
Tel.: +49 030 42 85 10 90
Fax: +49 030 42 85 10 92
INTERNET: http://www.logos-verlag.de

Preface

The work presented in this thesis has been carried out within the scope of my employment as a researcher at the Faserinstitut Bremen e.V. (FIBRE). The FIBRE is an independent research institute associated with the University of Bremen and specializes on fiber materials and fiber reinforced polymers. Part of this work has been carried out within the research project LoKosT (grant ID 20W0605A), with funding from the German Federal Ministry for Economic Affairs and Energy through the Federal Aeronautical Research Program IV (LuFo IV-2). The financial support through this project is gratefully acknowledged.

I want to thank my supervisor Prof. Dr. Axel S. Herrmann for offering me the position at the FIBRE and granting me the possibility and freedom to create this thesis. I also thank Prof. Dr. Hans-Günther Reimerdes for assuming the position as second examiner and the given advice and suggestions for completing this work. Further thanks go to Prof. Dr. Hans-Werner Zoch as chairmen of the examination board and Dr. Konstantin Schubert, Arne Breede and Bianca Reinhardt as board members.

I experienced my time at the FIBRE as both very encouraging and challenging and want to thank my colleagues, present and former, for their help and support. A special thanks goes to Dr. Christian Brauner for his general support of my work and numerous scientific discussions on composites and their mechanical response as well as Manuela von Salzen for performing many tests on my behalf. Furthermore I want to thank all project partners and team members of LoKosT for their support and endurance during sometimes prolonged discussions. Here I want to particularly mention Dr. Pierre Zahlen, Benjamin Teich, Gerrit Neumann, Johannes Eckhardt, Dr. Martin Rinker, Marianne John and Ali Yildiz.

In fall 2012 I stayed for three month in Göteborg, Sweden as a visiting researcher at Swerea SICOMP AB in the suburb Mölndal. I want to particularly thank Dr. Robin Olsson for his support of this stay and numerous discussions on the material and structural response of composite sandwich structures. Further thanks goes to Dr. Renaud Gutkin, Prof. Dr. Leif Asp and all other colleagues at Mölndal for welcoming me and their openness and willingness to share knowledge and experience with me.

Finally I want to thank my beloved wife Linda for her patience and continuous support during the time it took me to write this work and particularly during my stay in Göteborg. I also want to thank my parents and my brother Jörn for their advice and support that helped me conclude this thesis.

Abstract

Composite sandwich structures combine the stiffness and strength of fiber reinforced plastic face sheets with the high bending stiffness of sandwich structures. Local stiffeners such as stringers are thus often rendered superfluous resulting in great light weight construction properties. Low manufacturing costs can be achieved by utilization of vacuum infusion technology and polymer foam cores, which leads to numerous applications for e.g. marine structures and wind turbine blades, but has yet not found widespread use in the aeronautics, where honeycomb core structures are more established.

The impact response of composite foam core sandwich structures is in its individual steps complex, but can generally be characterized by two competing sandwich failure modes. Impact loads may cause sandwich failure due to face sheet rupture or core shear failure which are typically preceded by local face sheet delaminations and core crushing. Face sheet rupture leads to local but severe face sheet damage, while core shear failure typically leads to core cracks due to high shear stresses in the core. For simple geometries both failure modes can be described with analytical failure criteria. These criteria were compared against test results showing good agreement and allow the construction of failure mode maps as a guideline for designers.

Low velocity impact tests were performed on sandwich panels made of carbon fiber reinforced plastic (CFRP) face sheets and a Rohacell 71RIST polymethacrylimide foam core. Test results show that face sheet rupture generally leads to smaller damages and higher visibility while core shear failure leads to larger damage areas and lower visibility. Impact tests at -55 °C may also initiate vertical foam core cracks.

A numerical simulation model of the sandwich impact process was developed using the explicit finite element program LS-DYNA. This model allows the analysis of complex geometries and may thus be used for virtual testing but also sensitivity studies. The nonlinear material response of the CFRP face sheets and the foam core was validated stepwise along coupon and element levels of the building block test approach. Validation of the full model was then performed using impact tests on flat sandwich panels. Good agreement of experiments and simulation was achieved.

The sensitivity of impact parameters was finally investigated using analytical failure criteria and the developed simulation model with focus on the resulting failure mode. The single most important parameter that influences the sandwich impact response is the geometrical sandwich configuration as it largely determines the core shear load. Other relevant sandwich impact parameters are local boundary conditions and impactor size which both lead to higher core shear loads and thus promote core shear failure. This applies at a smaller degree also for thermal strains and impact velocity.

Keywords: Composite Materials, Sandwich Structures, Low Velocity Impact, Foam Core, Failure Mode, Core Shear Failure, Finite Element Analysis.

Kurzfassung

Sandwichstrukturen mit Deckschichten aus faserverstärkten Kunststoffen verbinden die Steifigkeit und Festigkeit eines Faserverbundwerkstoffs mit der hohen Biegesteifigkeit einer Sandwichstruktur. Lokale Versteifungsmaßnahmen der Struktur werden dadurch weitestgehend überflüssig, so dass Faserverbund-Sandwichstrukturen ein hohes Leichtbaupotential besitzen. Durch die Verwendung von Trockenfaserinfusionstechnologien und Schaumkernen können zudem geringe Fertigungskosten erzielt werden, was zahlreiche Anwendungen z.B. im Marineschiffbau und bei Rotorblättern für Windkraftanlagen zeigen. In der Luftfahrt werden hingegen fast ausschließlich Honigwabenstrukturen als Stützwerkstoff verwendet.

Das Impactverhalten von Sandwichstrukturen mit Schaumkern ist in seinen Einzelschritten sehr komplex, kann aber durch zwei miteinander konkurrierende Versagensmodi beschrieben werden. Impactlasten führen bei Sandwichstrukturen neben einer lokalen Stauchung des Schaumkerns und Delaminationen in der Deckschicht zum Versagen durch Deckschichtbruch oder einen Schubbruch des Schaumkerns. Deckschichtbruch und Kernschubversagen können bei geometrisch einfachen Problemen mit analytischen Versagenskriterien beschrieben werden. Die Versagenskriterien zeigen eine gute Übereinstimmung mit Versuchsergebnissen und wurden dazu genutzt, eine Dimensionierungshilfe für Sandwichstrukturen zu erstellen.

Impactversuche an Sandwichplatten aus kohlenstofffaserverstärktem Kunststoff und einem Rohacell 71RIST Polymethacrylimid Schaumkern zeigen, dass ein Deckschichtbruch zu einem gutmütigeren Strukturverhalten mit kleineren Schäden und guter Sichtbarkeit führt, während Kernschubversagen größere Schäden bei schlechterer Sichtbarkeit erzeugt. Impacts bei -55 °C können darüber hinaus zu vertikalen Brüchen im Schaumkern führen.

Eine numerische Simulation des Impactvorgangs wurde mit dem expliziten Finite Elemente Programm LS-DYNA aufgebaut. Das Simulationsmodell ermöglicht die Untersuchung einer komplexen Geometrie und kann daher für virtuelle Tests und Sensitivitätsanalysen eingesetzt werden. Das nicht-lineare Materialverhalten der Deckschichten und des Schaumkerns wurde schrittweise auf der Coupon- und Elementebene der Testpyramide validiert. Die Validierung des Gesamtmodells erfolgte mit Impactversuchen an Sandwichplatten und zeigte eine gute Übereinstimmung zwischen Experiment und Simulation.

Abschließend wurde die Sensitivität von Impactparametern auf den Versagensmodus mit den analytischen Versagenskriterien und dem numerischen Simulationsmodell untersucht. Die wichtigste bekannte Einflussgröße ist die Dicke des Sandwichkerns, da diese maßgeblich die Kernschublast bei Biegung bestimmt. Weitere relevante Einflussgrößen sind lokal die Lagerungs- und Randbedingungen sowie die Größe des Impactkörpers, welche Schublasten im Schaumkern und damit Kernschubversagen beeinflussen. Dies gilt in geringerem Maße auch für thermische Eigenspannungen und größere Aufprallgeschwindigkeiten.

Schlüsselwörter: Faserverbundwerkstoffe, Sandwich, Impact, Schaumkern, Versagensmodus, Kernschubbruch, Finite Elemente Analyse.

Table of contents

Analysis of the mechanical response of
impact loaded composite sandwich structures
with focus on foam core shear failure

Catalogue of symbols

Abbreviations, definitions

2D, 3D	two-dimensional, three-dimensional
3PB	three-point bending
AFRP	aramid fiber reinforced plastics
BBA	building block approach
BC	boundary condition
BVID	barely visible impact damage
CAI	compression after impact
CFRP	carbon fiber reinforced plastics
CLT	classical lamination theory
CSB	cracked sandwich beam
CTE	coefficient of thermal expansion
DLR	German Aerospace Center
EASA	European Aviation Safety Agency
FAA	Federal Aviation Administration
FE, FEM	finite element, finite element method
FIBRE	Faserinstitut Bremen e.V.
FRP	fiber reinforced plastics
FVC	fiber volume content
GFRP	glass fiber reinforced plastics
ILS	interlaminar shear
LEFM	linear elastic fracture mechanics
MMB	mixed mode bending
MVI	modified vacuum infusion
NCF	non-crimp fabric
NDI	non-destructive inspection
PEEK	polyether ether ketone
PET	polyethylene terephthalate
PMI	polymethacrylimide
PP	polypropylene
PPS	polyphenylene sulfide
prepreg	pre-impregnated, a class of semi-finished composite materials containing matrix and reinforcing fibers that are ready for curing
PS	polystyrene
PUR	polyurethane
PVC	polyvinyl chloride

RT	room temperature
SCB	single cantilever beam
TFC	tied foam core
VCCT	virtual crack closure technique
VID	visible impact damage
UD	uni-directional
WWFE	World Wide Failure Exercise

Symbols, Latin letters

A	shell membrane stiffness
A_L, A_p	area, subscript L denotes load introduction area, p plate area
a, a_m, a_p	core crushing radius, subscript denotes membrane (m) or plate (p) solution
a_d	delamination radius
a_1, a_2	material parameters of invariant failure criterion for PMI foams
$\bar{a}^2$	dimensionless core crushing radius, equivalent to the load fraction carried
B	stiffness of the coupling matrix
b	beam width
$b_{V,s}$, $b_{V,C}$	material shear (s) and compressive (c) strength ratio of PMI foam
$C_{A1}^{\pm}$	stiffness recovery factor of damaged UD plies in compression
C_B	adjustment parameter for stiffness recovery of damaged UD plies
C_{cell}	cell geometry factor for foams
C_{cz}	numerical cohesive zone length parameter
C_E, C_G, C_τ	combined cell geometry and material factors for foams
C_{rate}	strain rate scaling factor, material dependent
c	speed of sound
D	flexural rigidity of a general beam cross-section
D_0	additional beam bending stiffness due to loading off-axis from the centroid
D_f^*, D_p^*	effective bending stiffness of the sandwich face sheet (f) and plate (p)
d	damping coefficient
d_i	damage variable, i denotes damage type
E	elasticity modulus, also used for energy (index dependent)
E_c	elasticity modulus of the sandwich core
$E_{c,z}$	elasticity modulus of the sandwich core in the thickness direction
E_{cz}	elasticity modulus of the cohesive zone
E_f	elasticity modulus of the sandwich face sheets
$E_{f,x}$, $E_{f,y}$	elasticity modulus of the sandwich face sheets in the in-plane directions
E_{f1}, E_{f2}	elasticity modulus of the sandwich upper (1) and lower (2) face sheets
E_{bs}	energy stored by elastic bending and shear deformation

E_{imp}	impact energy
E_{kin}	kinetic energy
E_m	energy stored by elastic membrane deformation
E_γ	indentation energy
E^*	effective elasticity modulus of the indentation contact partners
E_f^*	effective elasticity modulus of the sandwich face sheet
E_r^*	averaged (quasi isotropic) elasticity modulus of the sandwich face sheet
E_{zc}^*	effective stiffness of the sandwich core
e	distance of face sheet centroid to sandwich centroid
F	force
F_L	applied load (force)
F_{cr}, F_{cs}	impactor force at core crushing (cr) or core shear failure (cs)
F_{dn}	impactor force necessary for growth of n face sheet delaminations
F_{dth}	impactor force at delamination initiation (delamination threshold)
F_{imp}, F_m, F_p	impactor force, subscript m denotes membrane theory and p plate theory
F_{max}	maximum impactor force
F_{rup}	impactor force at face sheet rupture
$\bar{F}$	normalized force
f_e	material effort
f_M, f_T	load factor for bending moment (M) and normal load (T)
G	shear modulus
G_c	shear modulus of the sandwich core
$G_{c,xz}$	shear modulus of the sandwich core in the xz-plane
$G_{f,xy}$	shear modulus of the sandwich face sheet in the xy-plane
G_{nc}	fracture energy (also: fracture toughness), n denotes opening mode I or II
G_{SL}	fracture energy of longitudinal shear UD ply failure
G_{tan}	tangential shear modulus of elasto-plastic UD ply shear response
G_{YC}, G_{YT}	compressive (C) or tensile (T) fracture energy of a UD ply, matrix direction
G_{XC}, G_{XT}	compressive (C) or tensile (T) fracture energy of a UD ply, fiber direction
G_{XC0}, G_{XT0}	compressive (C) or tensile (T) crack propagation fracture energy of a UD ply for use with bi-linear damage evolution, fiber direction
G^{free}	complimentary free energy
g	fracture toughness ratio
h	height (thickness)
h^*	relative thickness of the sandwich face sheet
h_c, h_{core}	sandwich core thickness
h_{cz}	cohesive zone thickness
h_d	distance between the centroids of the face sheets

h_f, h_{face}	sandwich face sheet thickness
h_{intf}	sandwich interface thickness
h_{ply}	layer or ply thickness
I	area moment of inertia
J	invariant of failure criterion for PMI foams
K_{cz}	cohesive zone stiffness
K_N, K_T	cohesive zone normal (N) and tangential (T) stiffness
K_{nc}	fracture intensity factor, n denotes opening mode I or II
K_z	elastic foundation modulus
k	spring stiffness
k_h	contact and indentation stiffness
k_b, k_m, k_s	plate bending (b), membrane (m) and shear (s) stiffness
k_{bs}	combined plate bending and shear stiffness
$\bar{k}_m$, $\bar{k}_{mcl}$	parameters of the dimensionless plate membrane stiffness
L	length
l_{el}	element length (in FEM)
M	bending moment
M'	first derivative of the bending moment
$\bar{M}$	normalized moment
m	mass
m_{imp}	impactor mass
m_p	plate mass
m_p^*	effective dynamic mass of the plate
$\bar{m}$	mass ratio of impactor and impacted plate
N	in-plane beam load
n	number (quantity)
n_{el}	number of elements across the cohesive zone
P	property of a foamed material
p	pressure
p_0	maximum impact pressure
p_{cr}	core constant reactive pressure equivalent to the core crushing / yield stress
Q	membrane stiffness in local coordinate system in plain strain conditions
$\bar{Q}$	membrane stiffness in global coordinate system in plain strain conditions
q	exponent of the contact stiffness relation
q_m, q_n	exponents of the foam core material strength (m) and stiffness (n)
R	radius
R_{cont}	contact radius
R^*	averaged radii of the indentation contact partners

r_N	material state variable, N denotes failure type
S, S_0	compliance, subscript 0 denotes undamaged state
S_L	longitudinal shear failure stress of a UD ply
S_τ	shear stress sensitivity parameter of impacted sandwich plates
S_f^*, S_p^*	effective shear stiffness of the sandwich face sheet (f) or plate (p)
s	dimensionless contact radius
T	transverse force
ΔT	thermal load
t	time
Δt	time step
$u, \dot{u}, \ddot{u}$	displacement, velocity and acceleration
v_0, v_{imp}	Initial velocity of the impact
W	work
w	face sheet displacement
X_C, X_T	compressive (C) and tensile (T) UD ply failure stresses, fiber direction
X_{C0}, X_{T0}	bi-linear compressive (C) and tensile (T) UD ply damage evolution stresses, fiber direction
Y_C, Y_T	compressive (C) and tensile (T) UD ply failure stresses, matrix direction
z_1	depth of the contact area between impactor and face sheet

Symbols, Greek letters

α_{cz}	scaling factor of the cohesive zone stiffness
α_T	coefficient of thermal expansion (CTE)
α_0	inclination of compressive matrix fracture plane
β_{cz}	cohesive zone coupling parameter of mode I and II failure
β_k	contact stiffness parameter
β_s	shear coupling parameter of fiber failure
Γ	ellipsoidal potential of PMI invariant failure criterion
γ	indentation of sandwich face sheet
γ_{cr}	indentation of sandwich face sheet that initiates core crushing
γ_0, γ_{0m}	indentation (0) and maximum indentation (0m) at the point of impact
Δ	cohesive zone displacement
Δ_n^0, Δ_n^F	cohesive zone displacement at damage initiation (0) and failure (F), n denotes opening mode I or II
ε	strain
$\varepsilon_C, \varepsilon_T$	ultimate compressive and tensile strain
ε_{SL}	ultimate UD ply shear strain for element erosion

ε_{XC}, ε_{XT}	ultimate UD ply compressive (C) and tensile (T) strains for element erosion, fiber direction
ε_{Y}	ultimate UD ply normal strain for element erosion, matrix direction
ε_{lu}	foam core lock-up strain
ε_{0}	face sheet membrane strain
ε_{1T}	laminate ultimate tensile strain
$\dot{\varepsilon}$	strain rate
$\dot{\varepsilon}_{ref}$	reference strain rate
ζ	exponent of membrane solution for sandwich indentation
η	stiffness knock down factor due to fiber waviness
θ	inclination of max. shear stress in the fracture plane
κ	curvature of a beam or plate
λ	wave length (of a harmonic wave)
μ_{L}, μ_{t}	friction coefficient of UD ply longitudinal (L) and transverse (t) to the fibers
ν	Poisson ratio
ν_{f}^{*}, ν_{p}^{*}	effective Poisson ratio of the sandwich face sheet (f) and plate (p)
ν_{r}^{*}	averaged (quasi-isotropic) Poisson ratio of the sandwich face sheet
ρ	material density
σ	normal stress
σ_{0}	face sheet membrane stress at the edge of the contact area
$\sigma_{f,x}$	face sheet normal stress in the x-direction
σ_{V}	equivalent normal stress
$\overline{\sigma}$	maximum normal stress
$\widetilde{\sigma}$	effective normal stress
$\widehat{\sigma}$	normal stress at material failure (also: ultimate normal stress)
$\widehat{\sigma}_{c,C}$	compressive yield stress of the sandwich core
$\widehat{\sigma}_{c,cr}$	crushing stress of the sandwich core
$\widehat{\sigma}_{c,cr}^{ref}$	quasi-static reference crushing stress of the sandwich core
$\widehat{\sigma}_{c,T}$	tensile yield stress of the sandwich core
τ	shear stress
τ_{c}	core shear stress
$\tau_{c,xz}$, $\tau_{c,yz}$	core shear stresses in the xz- and yz-planes
τ_{L}, τ_{t}	longitudinal (L) and transverse (t) shear stresses in the fracture plane
$\overline{\tau}$	maximum shear stress
$\overline{\tau}_{c}$	maximum core shear stress
$\tilde{\tau}$	effective shear stress
$\hat{\tau}$	shear stress at material failure (also: ultimate shear stress)
$\hat{\tau}_{c}$	core shear failure stress

$\hat{\tau}_Y$	UD ply shear failure stress
τ^0	interface strength
τ^{0*}	interface strength adjusted to a specific element size
τ_n^0	interface strength mode, n denotes opening mode I or II
φ^c	initial fiber misalignment
ϕ_N	failure criteria for UD ply which is also used as damage activation function, N denotes failure type
χ_1, χ_2	indentation parameters with geometric and material information
ψ_N	damage variable of UD ply, N denotes damage type
ω	impact frequency

Important subscripts

I, II	crack opening modes I (normal mode) and II (shearing mode)
b	bending
bs	combined bending and shear
C	compression
c, core	sandwich core
cr	core crushing
crit	critical, associated with initiation of damage or failure
cz	cohesive zone
el	element (in FEM simulations)
f, face	sandwich face sheet
fib	fiber direction of a UD ply
imp	impactor
intf	sandwich interface
L	longitudinal (fiber direction)
m	membrane
mat	matrix direction of a UD ply
N	failure or damage mode type
n	opening mode type
p	plate
ply	ply or layer (of a laminate)
r	radial
s	shear
sw	sandwich
u	unfoamed (solid) plastic material
T	tension
t	transverse (matrix direction)

Coordinate systems

1, 2, 3	local material coordinate system of a UD ply
x, y, z	laminate global and sandwich coordinate systems
r, ϑ, z	impactor cylindrical coordinate system

1 Introduction

Fiber reinforced plastics (FRP) are typically based on glass (GFRP), carbon (CFRP) or aramid fibers (AFRP) and a polymeric matrix and often simply referred to as composites or composite materials. The term composite or composite material is thus used as a synonym for fiber reinforced plastics throughout this work. Composite materials combine high specific stiffness and strength with superior fatigue properties making them ideally suited for light weight construction.

Sandwich structures are built of two macroscopically different materials and designed to provide superior bending stiffness and in-plane stability behavior at low specific weights. Consequently a combination of composite materials and sandwich structures to a composite sandwich structure is very attractive for light weight construction.

Sandwich structures, often misleadingly classified as a material type, are indeed small scale structures in themselves consisting of two or sometimes three macroscopically different materials that are permanently attached to each other. A sandwich structure is typically plate or shell type and consists of a thick core, which is engulfed by two thin but strong face sheets. The two different materials are permanently attached by different means of bonding, which forms the interface between the core and the face sheet as shown in Figure 1.1. The sandwich face sheets, which are the only visible part of a sandwich structure, are typically made of high performance structural materials. The core of a sandwich structure is made of lightweight materials and typically not visible from the exterior. Bonding of the face sheet to the core can be achieved by either secondary bonding or surplus resin of the applied composite system in the face sheets.

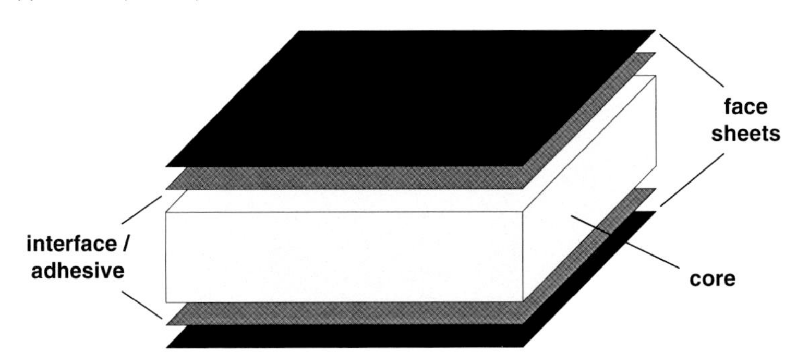

Figure 1.1 – principal elements of a sandwich structure

Numerous sandwich constructions are possible depending on the material selection of face sheet, core and interface and their geometric dimensions in the thickness direction. To distinguish between these, the term sandwich configuration will be used in this work. It describes the particular sandwich architecture applied including the selected face sheet and core materials as well as their thicknesses and material orientations. Furthermore the term composite sandwich structure shall describe a structure made of composite face sheets and a polymeric foam core. Industrial practice in material selection and manufacturing of sandwich structures differs significantly depending on the individual application. For this reason a brief overview will be given with the goal to explain differences and similarities and thus the scope and applicability of this work.

1.1 Motivation

Monolithic composite structures made of CFRP have, due to the high mechanical performance of the composite material itself, superior structural properties in terms of stiffness, strength and fatigue compared to metallic structures made of e.g. aluminum or steel. Disadvantages of composite structures are more complex design and manufacturing processes which lead to increased production costs. The higher production costs are mostly due to a low degree of automation, lower material output per hour and often higher raw material costs relative to metallic parts. Contributing to this is partially a lack of experience in the design of composite structures which may lead to so called "black metal" designs, which severely confine the application of composite structures and curtail their weight advantage with a high part count, required fasteners and assembly costs. This topic is currently addressed by several research projects that focus on improvements along the manufacturing chain of the typically prepreg based manufacturing process of composite structures for aerospace applications. An alternative approach to this is the development of simplified manufacturing processes based on less expensive textile semi finished products.

One more radical approach, which may induce a step change in production costs when compared to monolithic composite structures, is the development of large scale composite sandwich structures with reduced part count and assembly costs. The major advantage of sandwich structures is their inherently high bending stiffness which leads to improved stability behavior of the structure under in-plane compressive and shear loads enabling radically simplified designs. Disadvantages are a low out-of-plane shear stiffness which limits full exploitation of the high bending stiffness and greater structural complexity in the design of damage tolerant sandwich structures. This approach requires a manufacturing processes that enables the production of composite sandwich structures in the necessary quality and quantity but also a deep structural understanding and related robust design methods. The development of manufacturing processes and design methods for composite sandwich structures have to be performed in close collaboration as typical for composite materials.

1.2 Objective

This thesis investigates the low velocity impact behavior of composite foam core sandwich structures as an essential part of background knowledge required for damage tolerant design of the same. The mechanical response of an impact loaded sandwich structure generally consists of a superposition of a local indentation problem with the global structural response. The term superposition must be used with care in this context as nonlinear geometric and material responses prohibit the application of linear superposition.

This work focuses on sandwich failure modes as the resulting damages differ significantly. Impact loads on sandwich structures typically generate local face sheet delaminations and core crushing which is ultimately followed by face sheet rupture or core shear failure. Face sheet rupture is well known and has been described in the literature while core shear failure has so far been given only little attention. Impact generated foam core shear failure is difficult to detect during routine visual inspections and thus considered more critical to damage tolerance than face sheet rupture. In consequence methods that are capable to describe core shear failure due to low velocity impact are investigated with the goal to determine factors that promote or prevent this failure mode.

The damage tolerance of composite sandwich structures is often associated with interface crack growth under static or cyclic loading and the effect of the inflicted damage on the residual strength as studied by e.g. Berggreen [Ber04] for maritime structures and Rinker [Rin11] for aeronautical structures. Damage tolerance is also affected by the impact resistance of a structure as the resulting damage size must be tolerable and can be the starting point of a crack growth analysis. Vice versa interface crack growth affects the impact damage size and thus impact resistance as core shear failure typically leads to crack growth at the rear face sheet interface during the impact. Simulation and analysis of the sandwich impact response thus requires knowledge of the interface behavior.

Furthermore this thesis contributes to the general understanding of composite sandwich structures to deliver a basis for a more widespread utilization. The capabilities of available analytical methods are investigated with the purpose to describe the impact response. Experimental investigations and numerical simulations of low velocity impacts are performed and results compared with each other.

Enlargement of knowledge thus concentrates on two disciplines. On the one hand the material and structural behavior is investigated based on experimental work. This focuses on damage formation during low velocity blunt impact events addressing the mechanical response of composite sandwich structures. Impact testing addresses experimental mechanics supporting the development of standardized test methods. On the other hand analytical and numerical methods for the description of impact events are investigated, tailored to the specific problem, applied and results compared with test evidence. These methods are applied to investigate influencing parameters on the sandwich failure mode which addresses knowledge on the level of analysis methods and supports the development of a thorough understanding of the relationships and parameters that influence the sandwich impact response. Analytical investigations apply a stress based failure criterion of Olsson and Block [Ols13] which itself is based on a three step sandwich indentation model of Olsson [Ols02]. These criteria allow the construction of failure mode maps for face sheet rupture and core shear failure as a design guideline for engineers. The explicit finite element method (FEM), as provided by the software LS-DYNA, is applied for numerical simulation of low velocity impact on composite sandwich structures. This requires the development of an economic modeling approach and significant material testing for determining material properties and calibration of material models to the as built structure.

1.3 Thesis structure

This thesis is structured into seven chapters. The initial chapter provides an introduction into the work, its motivation, objective and structure.

Chapter two starts with a brief description of sandwich materials and manufacturing processes summarizing the state of the art for composite sandwich technology. Fundamentals of sandwich beam theory are explained as necessary for the understanding of analytical indentation and impact models of composite sandwich structures. Furthermore sandwich failure modes and failure mode maps are explained using the simpler sandwich beam problem to support understanding of the more complex response of a sandwich plate.

The third chapter initially describes the terms impact response, damage tolerance and building block test approach to define the scope of this work more precisely. It then provides a literature review of the impact response of composite sandwich structures and focuses on analytical methods for the description of the same. It concludes with the development of a failure mode map for impact loaded composite sandwich structures that distinguishes between foam core shear failure and face sheet rupture.

Experimental investigations of the sandwich impact response are covered by the fourth chapter. Test matrix, procedure and results are presented for impact tests at room temperature (RT) and -55°C. The test results include damage type and size as well as the impact response characterized by impact force vs. displacement data. Finally the observed damage is categorized and compared with the failure mode map of the previous chapter.

The fifth chapter describes the development of a numerical simulation model for impact on composite sandwich structures. As the impact response is highly nonlinear process, the stepwise building block test approach is adopted. Simulation approach and modeling techniques are discussed and applied to the specific problem. This includes theoretical background, constitutive material laws, numerical implementation and verification by test. Finally simulation results of low velocity impact on composite sandwich structures are presented and compared with test results of chapter four.

Influencing parameters on the impact response are investigated in the sixth chapter with focus on foam core shear failure. Analytical and numerical methods of chapters three and five are applied and results presented. Conclusions are drawn for the design of composite sandwich structures and experimental observations.

The seventh chapter summarizes the results of this work describing limitations and implications of the results. Recommendations for future work are given.

2 State of the art of composite sandwich technology

2.1 Application of composite sandwich structures

The systematic development of fiber reinforced plastics with manmade fibers and a polymeric matrix did not start until around 1940 when glass fiber as reinforcement and phenolic resins as matrix material became available [Dan06]. Subsequently more advanced reinforcement materials were developed leading to the availability of boron, carbon and aramid fibers as reinforcements. Similarly polymers as matrix evolved and lead to the development of advanced thermoset materials such as epoxy or vinyl ester resin systems or high performance thermoplastics.

The eatable sandwich is often referred back to John Montagu (1718-1792), 4th Earl of Sandwich and British lord of the Admiralty during the American Revolution [Zen97]. There are different tales about how the sandwich, which consists of two slices of bread which engulf a tasty filling such as e.g. meat or cheese, got his name. Sandwich structures are built similarly in that the slices of bread are replaced by slices of high-performance engineering material such as metal or composites that enclose a weak but light filling material. The slices of engineering material are now referred to as face sheets while the filling is the core of the sandwich, see also Figure 1.1. Early applications of sandwich structures were for a long time limited to steel bridge panels but evolved similarly to the development of structural adhesives and composite materials [Zen97].

One of the earliest applications of composite sandwich structures was the World War 2 era de Havilland Mosquito aircraft. Large parts of the structures were built of birch plywood face sheets bonded to a light balsa wood core with a phenolic resin system [Her05]. After World War 2 honeycomb cores made of aluminum and later aramid paper were developed and are extensively used in the aerospace sector.

In civil transport aircraft however composite sandwich structures have not found significant use in primary load bearing structures [Her05]. This is in contrast to their widespread use for secondary structures such as fairings, flaps, slats, rudders, engine cowlings and interior parts demonstrating their weight saving potential (Figure 2.1). The limited use in primary load bearing structures may be attributed to multiple reasons. One of these is their limited ability to withstand foreign object impact and related damage tolerance requirements.

Outside the aerospace industry sandwich structures have found large scale applications in the maritime and wind power industry as well as in vehicle construction. The maritime industry developed high performance racing boats and yachts made of GFRP or CFRP face sheets combined with foam cores. Also navy vessels are made of sandwich structures. Starting with mine hunters and patrol boats more recently also larger ships such as the Visby class corvette or deck super structures of frigates and destroyers are built employing composite sandwich structures [Zen97].

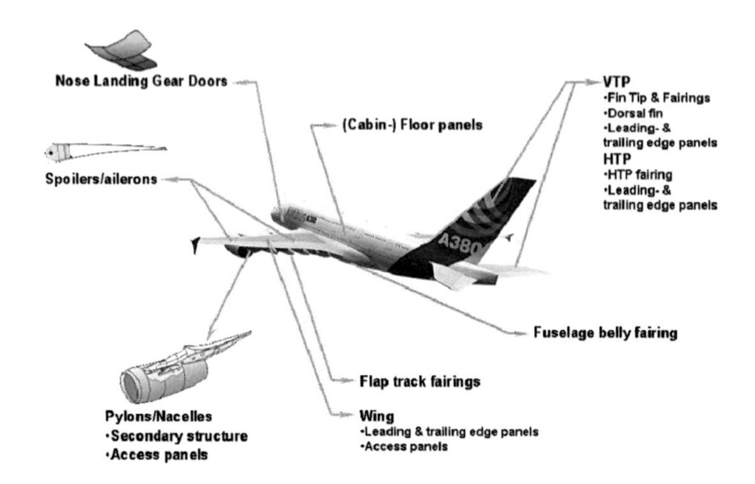

Figure 2.1 – sandwich structures in a modern civil transport aircraft (Airbus A380) [Her05]

The wind power industry is a large scale user of composite sandwich structures employing GFRP and various foam cores or balsa wood in the rotor blades of wind energy turbines but also in the nacelle and turbine housing. Figure 2.2 shows the cross-section of a typical rotor blade highlighting the extensive use of composite sandwich structures within the webs of the load carrying box girder and the aerodynamic profiles of the turbine blade.

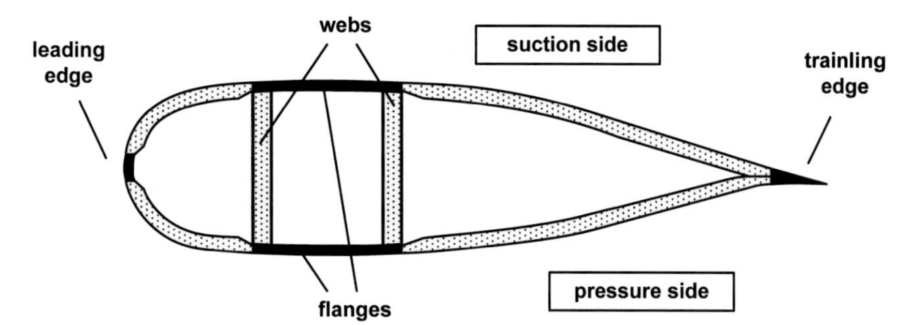

Figure 2.2 – typical cross-section of a wind energy rotor blade with a load carrying box girder; sandwich structures are indicated yellow

In vehicle construction composite sandwich structures are employed in applications such as refrigerated trucks and trailers with a combination of insulating and structural requirements. Also large scale application is found in the automotive interior. Structural applications may go as far as using steel face sheets with a solid plastics core such as polypropylene (PP). This allows lower weight and/or better acoustic insulation properties compared to a monolithic steel design [Thy11].

2.2 Materials and manufacturing

Face sheet materials

The principal purpose of the face sheet is to carry the in-plane and bending loads applied to the sandwich structure. Thus the face sheets are typically made of structural engineering materials such as metals or composites.

Fiber reinforced plastics are applied extensively in light weight construction including the face sheets of sandwich structures and itself composite materials made up of reinforcing fibers and a contour shaping matrix. Depending on the fiber, which dominates the mechanical properties of the composite, one distinguishes between glass, carbon or aramid fiber reinforced plastics. The function of the matrix material, which surrounds the fibers, is load introduction, macroscopic shaping of the part and protection of the fibers against environmental hazards. Mostly thermoset plastics such as epoxy, phenolic or vinyl ester resin systems are used while recently also thermoplastic matrix materials such as polyphenylene sulfide (PPS) or polyether ether ketone (PEEK) have been introduced.

CFRP with thermoset epoxy resin has found large application in the aerospace industry and has thus become a reference engineering material. This work focuses on CFRP face sheets only. Other available face sheet materials such as e.g. aluminum, steel, solid plastics or more unconventional materials such as wood are not considered.

Core materials

The principal purpose of the core is to increase the distance between the face sheets and take shear load and provide shear stiffness. This increases the bending stiffness and is the key to the superior mechanical performance of a sandwich structure compared to a monolithic design. Thermal and acoustic insulation functions can be integrated into the core.

Honeycomb, corrugated or folded cores are discontinues core structures. Honeycomb is best known and has often established as reference material in the aerospace industry. Here honeycomb cores are typically made of aluminum or aramid paper. Alternatives are steel or solid plastics. Honeycomb has very good weight vs. shear strength and stiffness properties and is – depending on the baseline material – thermally resistant up to 180°C. The open core cells limit however manufacturing processes and can lead to water accumulation in the core [Zen97].

Folded core structures are a more recent development with the goal to prevent water accumulation [Hei07, Grz10]. An origami alike continuous folding process is applied on prepreg CFRP or AFRP material. Variations in the folding process influence core characteristics such as thickness, stiffness, strength and density and offer a system of different cores [Hei08]. Corrugated cores are a way of producing discontinues cores more economically but are structurally less effective as honeycomb core materials [Zen97].

Foam cores were initially applied as a low cost solution in the maritime and wind power industries but also for gliders and have found widespread application today. Closed cell foam cores, especially polystyrene (PS), polyurethane (PUR), polyethylene terephthalate (PET) and polyvinyl chloride (PVC) foam, dominate while for aerospace applications high performance materials such as polymethacrylimide (PMI) foams are used [Zen97, Rot10]. At densities below 100 kg/m^3 foam cores typically have a lower mechanical performance than honeycomb. Depending on the baseline material foams are thermally stable up to 120 °C, PMI foams up to 180 °C. Foams can be used as a manufacturing aid. The small closed foam cells typically prevent water accumulation in the core.

Other common core materials are balsa or light ply wood and solid plastics. Balsa and light ply wood is typical for the maritime or wind power industries while solid plastics are more typical for automotive applications in combination with metallic face sheets [Thy11].

Interface

As a sandwich structures combines two macroscopically different materials, the interface between them is potentially a weak spot. This interface can be made of structural adhesives creating a secondary bonding, which is the case for most sandwich structures made of either discontinues cores (e.g. honeycomb) or non composite face sheets.

Composite face sheets allow the use of surplus resin from the manufacturing process as adhesive thus creating a co-bonding interface. This works best together with foams, balsa wood or similar closed cell core materials that have sufficient surface roughness providing a good link to the resin system. Also these core materials must not have open voids that fill with liquid resin.

Manufacturing processes

The manufacturing process of composite sandwich structures differs depending on the core type. For a better understanding, the fabrication of a honeycomb based and a foam core based composite sandwich structure shall be explained briefly.

Manufacturing of a honeycomb based composite sandwich structure requires a bonding process of the face sheets and the core. This can be achieved by co-bonding using an adhesive layer or a resin rich prepreg composite face sheet. As the prepreg material is often used uncured, the composite face sheet is in danger of building up fiber deformations due to the open cells of the honeycomb, see Figure 2.3 (a). Care must be taken with respect to the sandwich interface, as the bond area of core and face sheet is very small due to the open cell structure of the honeycomb. Liquid infusion of honeycomb cores requires an additional manufacturing step as the resin otherwise fills the open honeycomb cells, see Figure 2.3 (b). For sealing of the core a resin film or a resin rich prepreg is applied. This core assembly is then pre-cured until a semi-finished curing state of the resin is reached. This provides sufficient impermeability to stop the liquid infusion resin from penetrating into the honeycomb cells but still allow sufficient bonding of the infused dry textiles [Men12].

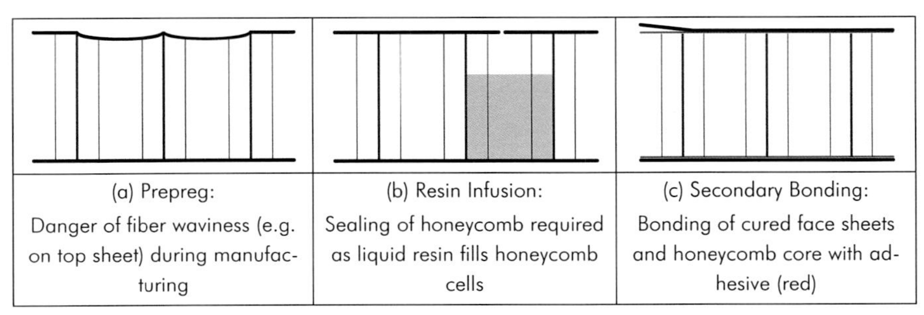

(a) Prepreg: Danger of fiber waviness (e.g. on top sheet) during manufacturing	(b) Resin Infusion: Sealing of honeycomb required as liquid resin fills honeycomb cells	(c) Secondary Bonding: Bonding of cured face sheets and honeycomb core with adhesive (red)

Figure 2.3 – manufacturing defects of composite honeycomb core sandwich structures

In consequence – alternative to the use of prepreg – a secondary bonding process must be used in combination with precuring of the face sheets. This leads to relatively high manufacturing costs due to two separate curing processes and related tolerances between the face sheet and the core. Attention must be paid to the surface quality which determines largely the quality of the bond, see Figure 2.3 (c). This often limits application of sandwich structures to non load bearing structures (secondary structures) as a secondary bonding process is currently not qualified for use in primary aircraft structures.

The manufacturing process of a composite foam core sandwich structure differs from this. As most structural foams have closed cells, liquid resin cannot penetrate into the core but will instead absorb resin in the interface to the face sheets. A comparison of the Rohacell foam types 71WF, 71RIST and 71RIMA showed that with decreasing cell size resin absorption of the core and thus the weight of the interface decreased, too [Evo13]. No exact values of the cell size are provided but these can be found in [Sae08]. Following this the foam type 71WF has with 0.743 mm the coarsest cell size, the foam type 71RIST with 0.295 mm an intermediate cell size and the foam type 71RIMA with 0.021 mm the by far smallest cell size. The sandwich peel strength – which can be used as a measure of the interface strength – stayed constant for the 71WF and 71RIST foams at around 7 MPa while it reduced to 3.5 MPa for the 71RIMA and thus only half the strength of the two other foams. This suggests that the intermediate cell size of the RIST foam is still large enough for maximum interface strength while further reductions of the cell size may lead to a trade-off between interface weight and strength.

Macroscopically the foam core supports textile preforms or uncured prepreg during manufacturing and thus acts as a tool. Prepreg or infusion processes which co-bond the yet uncured face sheets to the core are suitable and thus avoid some of the difficulties associated with the secondary bonding process that hampers honeycomb sandwich structures. Liquid resin infusion is mostly employed for composite foam core sandwich structures. The process emerged from the maritime and wind power industries where it is applied for large and complex structures. Here closed cell foam cores, especially PVC, PUR and PET foams, as well as balsa wood cores dominate the application while the face sheets are made of GFRP or CFRP depending on the structural performance required.

As a closed cell foam core is impermeable to the resin, it is effectively a barrier and forces the manufacturer to treat the impregnation of the two face sheets separately. Both impregnation processes are started by a single injection at the same time but two flow fronts emerge, one at the top face sheet and one at the bottom face sheet. As the sandwich is often built-up symmetrically, this does usually not create significant problems to the manufacturer but provides additional complexity compared to a monolithic part as demonstrated in Figure 2.4 [Zah08]. Problems related to two separate flow fronts can be alleviated by applying regularly spaced cuts in the core. This connects the flow fronts but introduces at the same time resin filled areas within the core adding weight and altering material properties. This is commonly applied for manufacturing of wind turbine blades.

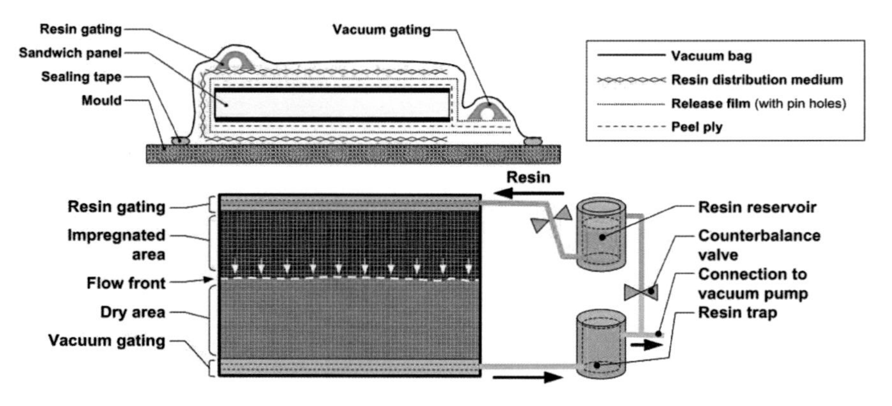

Figure 2.4 – vacuum assisted resin infusion of foam core composite sandwich parts [Zah08]

Reinforcements can be integrated into the foam core to improve material properties either locally at e.g. load introduction points or globally. In principal this is possible for all core materials and mostly applied with small metal plates or sleeves as local reinforcements for fasteners. Other reinforcements include regularly spaced pins as one-dimensional reinforcement or longitudinal profiles made of FRPs as two-dimensional reinforcements. These aim to improvements of core properties and/or provide a secondary load path. Here foam cores have the inherent advantage that they allow the use of dry textiles as base material for core reinforcements. During resin infusion the dry reinforcement are impregnated and create a co-cured composite within the form as shown in principle in Figure 2.5.

Pins as one-dimensional core reinforcements are very flexible. Different manufacturing technologies are available or subject to ongoing research. The company Aztex Inc developed the K-/X-Core™ technology based on pultruded carbon composite rods. Precured rods are pushed into the foam using force and ultrasonic vibration and connect with the face sheets during cure of the lamina [Mar05]. Due to the high price of pultruded rods textile technologies based on dry fiber rovings are economically more attractive. Of these the tied foam core (TFC) technology provides a high degree of flexibility by varying number and inclination of the pins [End10]. Here a thin fleece or fabric is temporarily fixed to both sides of the core. Next a needle stitches through the foam core, picks up a pre-cut roving and pulls it through the core leading to low fiber waviness. After infusion and co-curing of pins and face sheets the fleece or fabric provide a mechanical link of the pin.

Other reinforcement technologies apply thicker needles capable of penetrating fabrics. Two-sided stitching technologies provide higher manufacturing speeds at the cost of extra weight and limitations in the choice of reinforcing fibers. One example of this is the stitching technology of Acrosoma NV [Ver11], which works best with aramid fibers.

Alternatively to pins longitudinal profiles as two-dimensional reinforcements can be applied. These are based on dry textile raw materials and preformed with temporary fixation using binder. This provides sufficient stiffness for handling of the preform during manufacturing. The profile is then placed in between different parts of the foam core prior to infusion. After impregnation and curing the profiles become part of the sandwich structure as an integral reinforcement within the foam core. Different kinds of reinforcements can be combined within a single structure permitting the construction of large and complex sandwich panels as discussed by Zahlen et al. [Zah08] and shown in Figure 2.6.

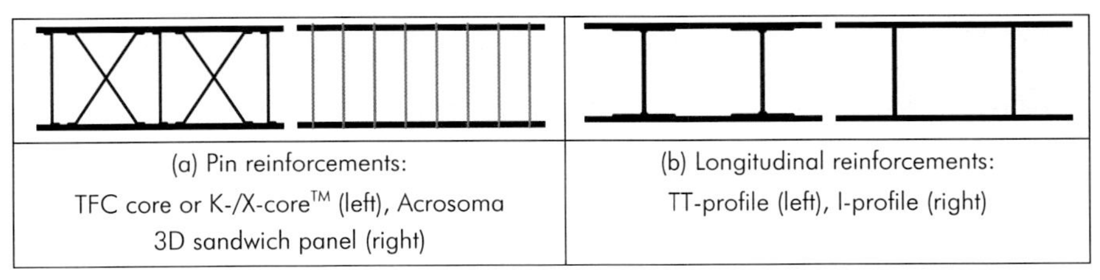

(a) Pin reinforcements: TFC core or K-/X-core™ (left), Acrosoma 3D sandwich panel (right)	(b) Longitudinal reinforcements: TT-profile (left), I-profile (right)

Figure 2.5 – foam core reinforcement types

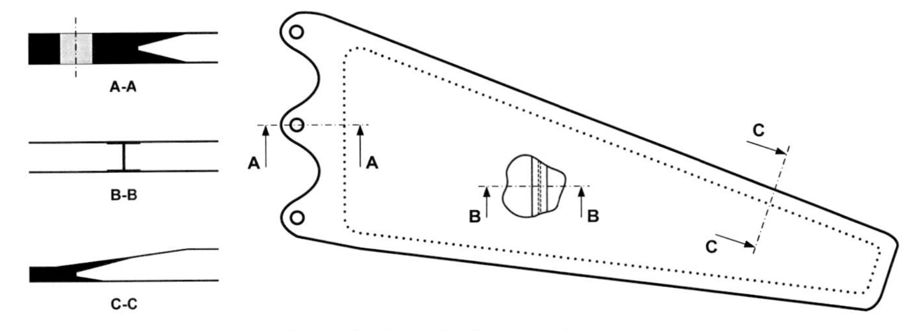

Figure 2.6 – complex sandwich panel with integrated foam core reinforcements

2.3 Fundamentals of composite sandwich theory

Analysis of the mechanics of composite materials led to the development of the classical lamination theory (CLT) and failure models. Starting in the beginning of the 20th century this development still continues with focus on the strength and failure behavior as discussed in section 5.2. For fundamentals on composite materials the reader shall refer to common literature such as Niederstadt [Nie97], Daniel [Dan06] or Schürmann [Sch07].

Historically research about sandwich structures dates back to the 1940's even though specific advantages of the sandwich concept were already discussed as early as in 1820 by Duleau and later by Fairbairn [Zen95]. As during the time of World War Two manufacturing technologies and consequently applications for sandwich structures evolved, research had to cope with this and led to more profound knowledge. A comprehensive introduction into the analysis of sandwich structures has been edited by Zenkert [Zen95] who refers to works of Allen [All69] and Plantema [Pla66]. A handbook for engineers that focuses on applications of sandwich structures has also been edited by Zenkert [Zen97]. An alternative reference is the work of Altenbach et al. [Alt96]. An extract of sandwich beam and plate theory based on the work of Zenkert [Zen95] is given in appendix A.

To illustrate the benefit of a sandwich compared to a monolithic structure one has to focus on the bending stiffness. The bending stiffness D of a sandwich beam as shown in Figure 2.7 compromises of the bending stiffnesses of the individual constituents face sheet (index f) and core (index c) in addition to the effect of the Huygens-Steiner theorem (index 0). The same indices are used to identify the geometry and material properties of the sandwich constituents. The resulting stiffness is described in equation (A.9) of appendix A

$$\mathrm{D} = \int \mathrm{E}z^2 \mathrm{d}z = \frac{\mathrm{E_f h_f^3}}{6} + \frac{\mathrm{E_f h_f h_d^2}}{2} + \frac{\mathrm{E_c h_c^3}}{12} = 2\mathrm{D_f} + \mathrm{D_0} + \mathrm{D_c}\ . \tag{2.1}$$

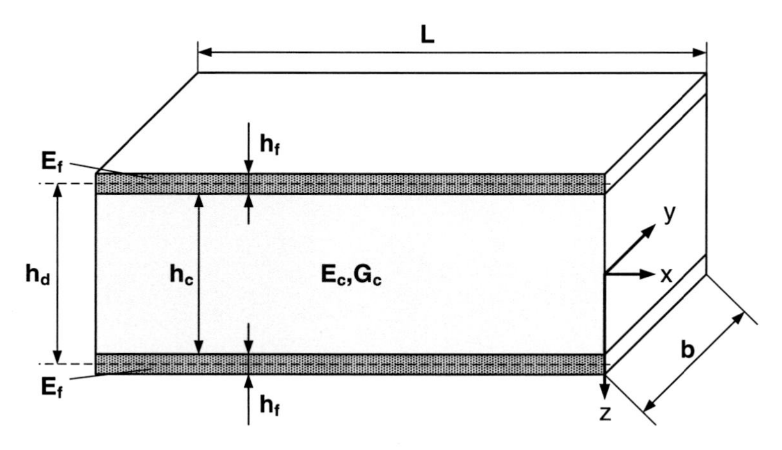

Figure 2.7 – sandwich beam properties and coordinates

Here the coordinates x, y and z specify the sandwich beam length, width and thickness while E_f and E_c describe the elasticity modulus of the face sheet and core. For sandwich beams with thin face sheets and a weak core this can be simplified (refer to equations (A.10) to (A.12) of appendix A). Using these assumptions the sandwich bending stiffness D reduces to D_0 which solely describes the contribution of the fact that the face sheets are placed outside of the bending axis as specified by the Huygens-Steiner theorem

$$D \approx D_0 = \frac{E_f h_f h_d^2}{2} \,. \tag{2.2}$$

The stresses in a sandwich beam due to global bending distribute between the face sheets and the core similar to an I-beam. The flanges, here the face sheets, carry in-plane tensile and compressive loads while the web, here the core, carries shear loads. Equation (A.21) of appendix A now applies the assumptions of thin face sheets and a weak core. Using M_x for the applied bending moment and $T_x = dM_x/dx$ for the change of the bending moment in the beam direction, the sandwich stresses can be described by

$$\begin{aligned} &\sigma_c(z) = 0\,, && \sigma_f(z) = \pm\frac{M_x}{h_f\,h_d}\,, \\ &\tau_c(z) = \frac{T_x}{h_d}\,, \text{ and} && \tau_f(z) = 0\,. \end{aligned} \tag{2.3}$$

Zenkert [Zen95] illustrates the sandwich effect by comparing the bending strength and stiffness of a monolithic beam with two sandwich beams where the face sheets are together as thick as the monolithic beam as shown in Figure 2.8. It is assumed that the bending stiffness and strength can be calculated as summarized in equations (2.2) and (2.3).

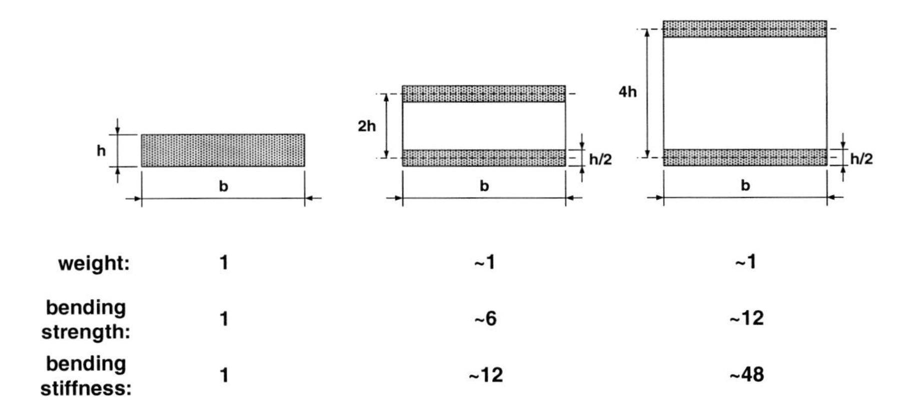

Figure 2.8 – sandwich effect as illustrated by Zenkert [Zen95]

Highly nonlinear stress and strain behavior in sandwich structures e.g. due to ply drop-offs can be described using higher-order sandwich beam and plate theories such as those of Frostig et al. [Fro92] and Thomsen [Tho00].

2.3.1 Localized loads on sandwich beams

Localized out-of-plane loads in sandwich beams or structures are typically occurring in load introduction areas or due to foreign object impacts and may cause a local indentation. This local effect adds to the stresses in the sandwich due to global loads. Considering e.g. the three-point bending (3PB) load case, the mechanical response can be determined by superposition of beam bending and local indentation as shown in Figure 2.9 [Lim04]. Superposition must be used with care in this context as nonlinear material and geometry behavior limit the applicability of linear superposition. Shipsha et al. [Shi03, Shi03a] perform impact tests on sandwich beams and determine their residual strength properties when subject to compression, shear and bending.

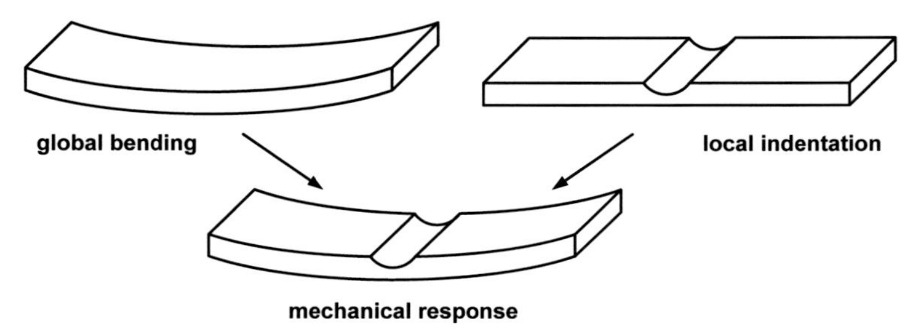

Figure 2.9 – mechanical response of a sandwich beam subject to bending [Lim04]

A simple model for calculation of localized indentation of sandwich beams and plates is referred to as the Winkler elastic foundation model [Het46]. Here the core provides locally only support to compressive loads and is thus modeled as continuously distributed tension and compression springs. Thomsen [Tho92, Tho95] summarizes in Zenkert's book [Zen95] a more elaborate two parameter model, which also accounts for shear loads between the face sheet and a linear elastic core material. The elastic response of the face is assumed as a harmonic wave. Its wave length λ can be described as

$$\lambda = \frac{2\pi}{\beta_k} = 2\pi \sqrt[4]{\frac{4\,D_f}{K_z}}\,. \tag{2.4}$$

Here K_z is the elastic foundation modulus of the face sheet provided by the core and D_f is the face sheet bending stiffness summarized by the contact stiffness parameter β_k. Depending on the relative face sheet thickness, the elastic foundation modulus becomes [Zen95]

$$\begin{aligned} K_z &= 0.28\,E_c \sqrt[3]{\frac{E_c}{D_f}} && \text{for} \quad h_f \ll h_c \quad \text{and} \\ K_z &= \frac{E_c}{h_c} && \text{for} \quad h_f \approx h_c\,. \end{aligned} \tag{2.5}$$

The limit of $h_f \ll h_c$ is not immediately clear and thus which equation to use for which values of h_f and h_c. Thus K_z is calculated for a number of reference cases using the material properties of the CFRP laminate and the PMI foam Rohacell RIST with the grades 51, 71 and 110 as summarized in appendix B. Here the relative face sheet thickness $h^* = h_f/h_c$ is used as characteristic value. The assumption $h_f \approx h_c$ becomes valid for sandwich structures with a face sheet thickness of $h^* \geq 0.1$ based on these results. Using the formulation of K_z for $h_f \ll h_c$, Thomsen provides as a guideline

$$\lambda \approx 5.33\, h_f \sqrt[3]{\frac{E_f}{E_c}} < 50\text{-}60 \text{ mm}. \quad (2.6)$$

If the wave length λ is below 50 - 60 mm, Thomsen recommends the use of the two parameter model. Only Indentations characterized by a short wave length λ are subject to significant shear interaction and thus require the more elaborate two parameter approach.

Using an in-plane stiffness of 50000 MPa typical for quasi-isotropic CFRP, a stiffness of 105 MPa for Rohacell 71RIST and a face sheet thickness of $h_f = 1.5$ mm λ becomes

$$\lambda \approx 5.33 * 1.5\text{mm} \sqrt[3]{\frac{50000}{105}} \approx \underline{62.43 \text{ mm}} > 60 \text{ mm}.$$

As the selected parameters are exemplary for the specimens used in this work, focus will be put on the Winkler elastic foundation model.

It should be noted, that λ is directly proportional to the thickness h_f of the face sheet. Thus a change in the face sheet thickness has a direct effect on the amount of shear that influences the local indentation behavior. Sandwich beams made of this material combination and face sheets with a thickness of $h_f < 1.5$ mm will have a stronger shear influence and thus require the more complex two parameter model outlined by Thomsen.

The Winkler elastic foundation model is based on the idea that the elastic response of the supporting medium – in this case the sandwich core – can be expressed by [Lim04]

$$q_z(x) = -K_z w_{local}(x) \ . \quad (2.7)$$

Here $w_{local}(x)$ is the local deflection of the loading point on the sandwich face sheet. Total deflection of the loading point can be described using superposition of the deflection due to global bending of the sandwich beam and indentation of the affected face sheet due to local face sheet bending and core indentation [Zen97]:

$$w_{total}(x) = w_{global}(x) + w_{local}(x) \ . \quad (2.8)$$

For determination of $w_{total}(x)$ the two parts $w_{global}(x)$ and $w_{local}(x)$ are determined separately. $w_{global}(x)$ can be calculated using ordinary sandwich beam theory while for $w_{local}(x)$ the loaded face sheet is modeled as an ordinary beam. By applying equilibrium to an infinitesimally small piece of the face sheet and determining equilibrium of all applied forces an ordinary differential equation of 4th order can be established with $w_{local}(x)$ as the unknown variable. For more details on the solution refer to [Zen97].

From this solution Lim et al. [Lim04] determine the maximum indentation δ_{local} of the loaded face sheet of a sandwich beam during 3PB to

$$\delta_{local} = \frac{F_L\,\beta_k}{2\,K_z} \qquad\qquad (2.9)$$
$$\text{with} \quad \beta_k = \sqrt[4]{\frac{K_z}{4D_f}} \quad \text{and} \quad D_f = \frac{E_f h_f^3}{12}\,.$$

The parameter β_k is already known from equation (2.4) where its inverse is used. Here F_L is the centrally applied load per unit width as shown in Figure 2.10. The stress field in the affected sandwich parts can be described similarly using superposition with

$$\sigma_{f,total}(x) = \sigma_{f,global}(x) + \sigma_{f,local}(x)\,,$$
$$\tau_{int,total}(x) = \tau_{int,global}(x) + \tau_{int,local}(x) \quad \text{and} \qquad (2.10)$$
$$\sigma_{int,total}(x) = \sigma_{int,local}(x)\,.$$

Here σ_f is the normal stress in the face sheets, τ_{int} the shear stress and σ_{int} the normal stress in the interface of the core to the loaded face sheet.

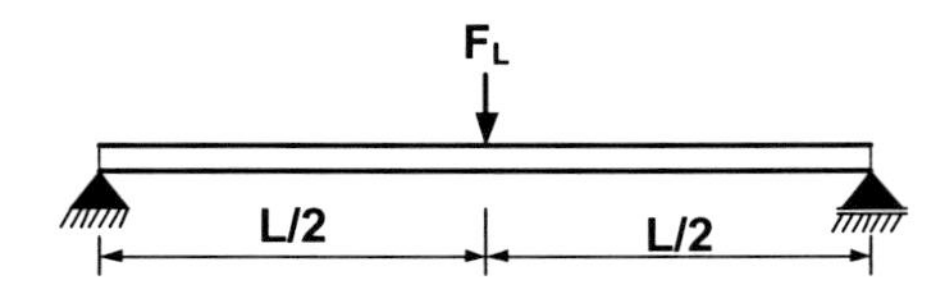

Figure 2.10 – three-point bending (3PB) of a sandwich beam

Only the two parameter model by Thomsen is capable of describing the influence of local shear deformation. Therefore $\tau_{int,local}$ is 0 for the simpler Winkler model. The maximum stress in the face sheet due to local bending is described by [Lim04]

$$\sigma_{f,local} = \frac{M_{local}}{D_f}\frac{h_f}{2}E_f = \frac{6\,F_L}{4\,h_f\,\beta_k} \qquad\qquad (2.11)$$
$$\text{with} \quad M_{local} = \frac{F_L}{4\beta_k}\,.$$

The global stresses can be determined using classical sandwich beam theory as outlined in appendix A.1. Equations (A.20) and (A.21) of appendix A are now used to express the maximum face sheet stress due to global bending [Lim04]

$$\sigma_{f,global} = \frac{M_{global}}{D}\frac{h_d}{2}E_f = \frac{f_M F_L L}{h_d\,h_f} \qquad\qquad (2.12)$$
$$\text{with} \quad M_{global} = f_M F_L L \quad \text{and} \quad D = \frac{E_f h_f h_d^2}{2}\,.$$

The total stress in the face sheet thus amounts to

$$\sigma_{f,total} = \sigma_{f,global} + \sigma_{f,local} = \frac{M_{global}}{D}\frac{h_d}{2}E_f + \frac{M_{local}}{D_f}\frac{h_f}{2}E_f$$
$$= \frac{f_M F_L L}{h_d\, h_f} + \frac{6\, F_L}{4\, h_f\, \beta_k} = P\left(\frac{f_M L}{h_d\, h_f} + \frac{6}{4\, h_f\, \beta_k}\right). \tag{2.13}$$

Using equation (2.7) and the width of the beam b, the compressive stress in the interface that acts on the core is described by

$$\sigma_{int,local}(x) = \frac{q_z(x)}{b} = \frac{-K_z w(x)}{b}. \tag{2.14}$$

2.3.2 Failure modes of sandwich beams

Sandwich beams are representative of more complex sandwich structures such as plates or shells. Sandwich structures may fail in several ways each representing individual failure modes. Every failure mode adds one constraint to the structure and thus has to be accounted for during design. Zenkert [Zen95] summarizes the most important sandwich failure modes as shown in Figure 2.11.

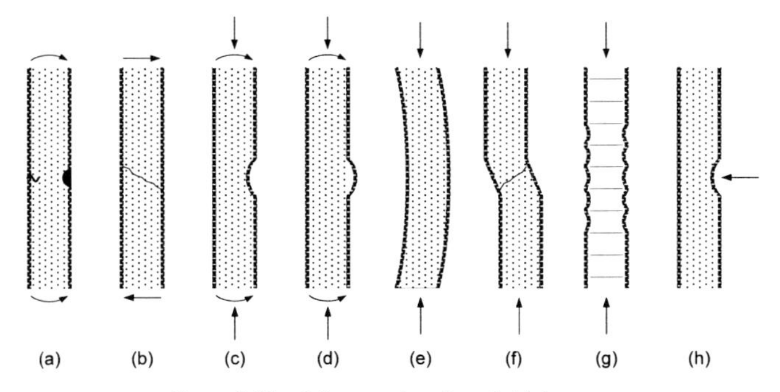

Figure 2.11 – failure modes of sandwich beams:
(a) face yielding/fracture, (b) core shear, (c) + (d) face wrinkling, (e) global buckling, (f) shear crimping, (g) face dimpling and (h) local indentation [Zen95]

For the design of a sandwich structure each failure mode has to be checked against all relevant load cases. The failure mode that occurs first then becomes the critical design case. Based on Zenkert's [Zen95] work sandwich failure modes will be discussed briefly with respect to their importance to the impact response of composite sandwich structures.

(a) In-plane yielding / fracture of the face sheets

Composites face sheets subject to an in-plane overload typically fail due to fracture both in tension or compression. As the fracture process is very complex, various failure criteria for fiber reinforced composites have been developed and will be discussed further in section 5.2. Metallic face sheets may also yield. In any case there is a maximum in-plane stress the face sheet can sustain. Face sheet failure due to in-plane stresses originating from global bending of a sandwich beam can be expressed using equation (A.20) by

$$\overline{\sigma}_{f,x} = \frac{M_x z E_{f,x}}{D_x} \geq \widehat{\sigma}_{f,x} \quad \text{with} \quad D_x = D_0 + 2D_f \,. \tag{2.15}$$

Here $\overline{\sigma}_{f,x}$ is the maximum face sheet stress while $\widehat{\sigma}_{f,x}$ is its ultimate stress which is equivalent to the material strength. This criterion has to be used for both face sheets as the material may have different ultimate stresses in tension and compression. As most core materials sustain significantly higher strain levels than the face sheets, the formulation of such a criterion for the core does not make sense. For in-plane loads of the sandwich beam the failure criterion reduces to

$$\overline{\sigma}_{f,x} = \frac{F_x}{2h_f b} \geq \widehat{\sigma}_{f,x} \,. \tag{2.16}$$

Here $\overline{\sigma}_{fx}$ is the maximum normal in-plane load in the sandwich beam due to bending and can be determined from equation (A.21).

(b) Core shear failure

As shown in Figure A.5 the core of a sandwich beam, which is subject to global bending, is predominantly loaded in shear. This load may now be superposed by normal stresses in the core due to in-plane loads of the sandwich beam using Mohr's circle of stress. The resulting failure criterion is expressed by Zenkert as

$$\overline{\tau}_{c,xz} = \sqrt{\left(\frac{\sigma_{c,x}}{2}\right)^2 + \tau_{c,xz}^2} \,. \tag{2.17}$$

Using the weak core assumption $E_f \gg E_c$ the normal stresses in the core can be neglected and equation (2.17) reduces to

$$\overline{\tau}_{c,xz} = \hat{\tau}_{c,xz} \,. \tag{2.18}$$

Here $\overline{\tau}_{cxz}$ is the maximum shear load due to bending and can be determined from equation (A.21) while $\hat{\tau}_{c,xz}$ is the ultimate stress and thus shear strength of the core material.

(c) + (d) Face sheet wrinkling

Face sheet wrinkling can occur in a sandwich beam due to in-plane compressive loads on the affected face sheets. This is the case when the sandwich beam is either subject to in-plane compressive loads or global bending. In case of global bending only the compressively loaded face sheet may be subject to wrinkling. The critical wrinkling stress is according to Zenkert

$$\sigma_{f,x} = \frac{1}{2}\sqrt[3]{E_{f,x} E_{c,x} G_{c,xz}} \,. \tag{2.19}$$

Zenkert states that this equation serves well for most practical cases. As impact loads cause bending loads on the sandwich, wrinkling failure may occur.

(e) General buckling

Buckling is a stability type failure mode and may cause a sandwich structure to either fail structurally or lose its ability to fulfill all functions without actually causing structural failure. As impact loads typically do not introduce in-plane compressive loads, general buckling will not be further discussed. It is however noted that impact damage may have a great effect on the buckling behavior of sandwich structures and thus should be considered when investigating the stability and damage tolerance of sandwich structures.

(f) Shear crimping

Shear crimping is also a stability type failure mode and driven by the sandwich buckling stability. The failure mode occurs predominantly due to in-plane compressive loads and is characterized by shear failure of the core often followed by immediate face sheet failure. Large out-of-plane deformations and a low core shear stiffness contribute to shear crimping. Failure criteria are typically formulated as limit case for general buckling using the thin face sheet assumption.

(g) Face dimpling

Face dimpling is a local stability failure mode that may occurs in discontinues core materials such as honeycomb. Here the face sheet locally buckles between the support of the core. The core itself is typically not damaged. As this work investigates only foam cores with a continuous support, face sheet dimpling is not an applicable failure mode.

(h) Core indentation

Core indentation occurs as a concentrated out-of-plane load acts on the sandwich such as e.g. in load introduction areas or due to impact loads. Thus load introductions have to be applied on an area A_L large enough which can be estimated from [Tri87]:

$$A_L = \frac{F_L}{\hat{\sigma}_c} \,. \tag{2.20}$$

Here $\hat{\sigma}_c$ is the compressive strength of the core material in sandwich thickness direction.

Impact loads act – depending on the type of the impactor – rather like a point load than an area load. Thus equation (2.20) is not applicable. Instead a point load causes the affected face sheet to locally bend like a plate and thus independently of the unaffected rear face sheet while the core acts like an elastic foundation. Once the compressive stress of the core is reached, the core will yield and thus fail. Using the Winkler elastic foundation model, the stress field in the beam can be described by superposition of the global and local solutions as shown in section 2.3.1. Based on this a failure criterion for core compression has been formulated by Lim et al. [Lim04]

$$\sigma_c = \frac{\delta_{local}}{h_c} E_c \geq \hat{\sigma}_c \,. \tag{2.21}$$

2.3.3 Failure mode maps

Using the previously discussed criteria for sandwich failure modes, one can now establish a failure mode map for 3PB of sandwich beams. This is a graph with relevant design parameters of the sandwich beam on the horizontal and vertical axes that plots equilibrium lines of each two failure modes. At first failure is divided into core shear failure and face sheet failure. Face sheet failure itself is further divided into the modes of yielding/fracture and wrinkling. Such a failure mode map has first been developed by Triantafillou and Gibson [Tri87] for optimal design of sandwich beams made of aluminum face sheets and polyurethane foam core subject to 3PB, see Figure 2.12.

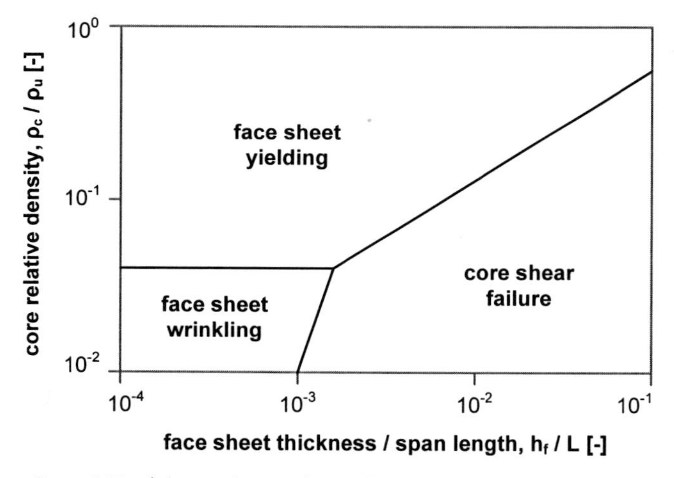

Figure 2.12 – failure mode map of a PUR foam core sandwich beam [Tri87]

Triantafillou and Gibson state that first one has to define the principal types of failure modes. Face yield or fracture and core shear are independent and certainly one of the most important failure modes of sandwich beams. Concentrating on transversal loads and thus neglecting in-plane compression as a load case, global buckling or shear crimping will not occur. Face dimpling is not applicable for foam cores and will thus be neglected here. Thus only local indentation and face sheet wrinkling on the compressive side are applicable. Avoiding local indentation by a sufficiently large load introduction area, Triantafillou and Gibson concentrate on face sheet wrinkling as the third independent failure mode. They also use the core density as a variable which provides the core properties.

Zenkert [Zen95] describes this failure mode map with a simpler notation based on the idea that foam cores are a three-dimensional network of struts, which react to loads by bending. This leads to a nonlinear relationship between the properties of the foam and its relative density. Gibson and Ashby derived this relationship for a large number of foam materials [Gib82, Gib88]. Generally, the properties of foam can be expressed by

$$\frac{P_c}{P_u} = C_{cell} \left(\frac{\rho_c}{\rho_u}\right)^n . \tag{2.22}$$

Here P_c is the material property of the foam, P_u the same property of the unfoamed and thus solid material, C_{cell} a constant related to the cell geometry, ρ_c and ρ_u the densities of the foam and the solid but unexpanded core material and n a constant. The advantage of this formulation is that the material properties of the foam core can now be expressed in terms of the density of the foam. Li et al. [Li00] published limited material properties for the PMI baseline material:

$$E_{PMI} = 5200 \text{ MPa}, \quad \text{and} \quad \hat{\sigma}_{PMI} = 90 \text{ MPa} .$$

Zenkert [Zen95] rearranges equation (2.22). Every material property of the foam core can then be expressed by

$$P_c = C_{cell}\, P_u \left(\frac{\rho_c}{\rho_u}\right)^{q_n} . \tag{2.23}$$

The product $C_{cell}\, P_u$ is now summarized in a single value:

$$E_c = C_E \left(\frac{\rho_c}{\rho_u}\right)^{q_n}, \qquad G_c = C_G \left(\frac{\rho_c}{\rho_u}\right)^{q_n}, \qquad \text{and} \quad \hat{\tau}_c = C_\tau \left(\frac{\rho_c}{\rho_u}\right)^{q_m} . \tag{2.24}$$

The values C_E, C_G, and C_τ and the exponents q_n and q_m can be extracted by curve fits from experimental results or published material data of the supplier. In case of the Rohacell RIST PMI foam these values were extracted using a curve fit of the published material data [Evo13]. It turned out that the curve fit works better if the given value for C_E is ignored and instead a best fit value used. In conclusion the following parameters for describing the material properties of the Rohacell RIST PIM foam core where extracted:

$$E_c = 4500 \text{ MPa} \left(\frac{\rho_c}{\rho_u}\right)^{1.35}, \qquad G_c = 1750 \text{ MPa} \left(\frac{\rho_c}{\rho_u}\right)^{1.35}, \qquad \text{and}$$
$$\hat{\tau}_c = 85 \text{ MPa} \left(\frac{\rho_c}{\rho_u}\right)^{1.5}.$$

To define a general load case Zenkert [Zen95] uses a modified notation to describe the maximum bending moment and transverse loads of a sandwich beam by

$$|M|_{max} = f_M F_L L , \quad \text{and} \quad |T|_{max} = f_T F_L . \tag{2.25}$$

Here F_L is the applied load and the constants f_M and f_T load factors depending on the load case which have to be determined for each case of interest separately. For 3PB as shown in Figure 2.10, the parameters are $f_M = 1/4$ and $f_T = 1/2$. Using equation (2.24) and (2.25) the expressions from equations (2.16), (2.18), and (2.19) are now rewritten to

Face yield: $$F_L = \frac{\hat{\sigma}_f h_f h_d}{f_M L},\tag{2.26}$$

Face wrinkling: $$F_L = \frac{h_f h_d}{f_T} \sqrt[3]{E_f C_E C_G \rho_c^{2n}}\text{ , and}\tag{2.27}$$

Core shear: $$F_L = \frac{C_\tau \rho_c^{\varrho} h_d}{f_T}.\tag{2.28}$$

For creation of a damage mode map, the transition lines between these failure modes have to be determined. This can easily be done by equating each two loads from equations (2.26) to (2.28). The resulting expressions describe transition between the failure modes. They are then solved for the core density ρ_c as relevant design parameter and the expression h_f/L, which is used on the horizontal axis. Failure mode transition is now expressed by:

(a) transition between face fracture and core shear failure

$$\rho_c = \sqrt[m]{\frac{f_T \hat{\sigma}_f}{f_M C_\tau}\left(\frac{h_f}{L}\right)},\tag{2.29}$$

(b) transition between wrinkling and core shear failure

$$\rho_c = \sqrt[(2n-3m)]{\frac{1}{E_f C_E C_G}\left(\frac{2\, f_M C_\tau}{f_T\left(\frac{h_f}{L}\right)}\right)^3}\text{ , and}\tag{2.30}$$

(c) transition between wrinkling and face sheet fracture

$$\rho_c = \sqrt[2n]{\frac{8\,\hat{\sigma}_f{}^3}{E_f C_E C_G}}.\tag{2.31}$$

Figure 2.13 now shows the damage mode map for a sandwich beam with a Rohacell RIST foam core and quasi-isotropic CFRP face sheets based on equations (2.29) to (2.31)

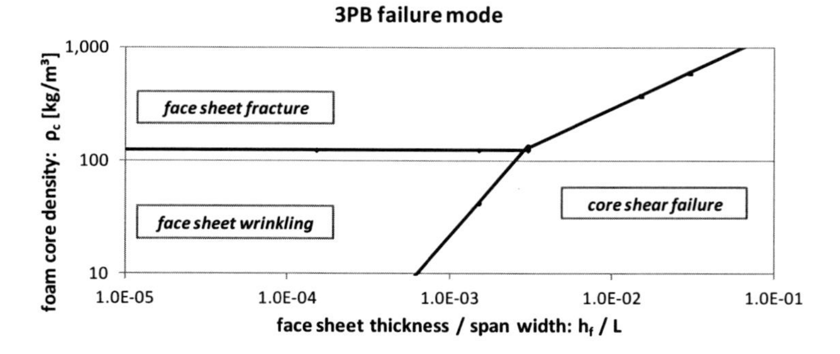

Figure 2.13 – three-point bending (3PB) failure mode map of a sandwich beam with a Rohacell RIST foam core with variable core density

The damage mode maps show that sandwich beams with a low core density tend to fail by face sheet wrinkling rather than face sheet fracture. Short beams with a small h_f/L ratio are loaded predominantly in shear while longer beams rather in the face sheets. Increasing the core density reduces this effect but does not eliminate it entirely.

Lim et al. [Lim04] design a failure mode map for three-point bending of sandwich beams in a similar way but use instead the parameters face sheet thickness h_f and span width of the beam L both relative to the core thickness h_c. Thus the properties of the selected materials are not variable in this failure mode map shown in Figure 2.14 for the same material combination of CFRP and PMI foam as used in the previous example. Instead the core thickness becomes a variable. This failure mode map instead focuses on geometric parameters only.

The failure modes considered by Lim et al. [Lim04] are also different. Face sheet fracture and core shear failure are both included, while the third investigated mode is core compression. For describing the later failure mode. Lim et al. use a superposition of global bending of the sandwich beam and local bending of the loaded face sheet as explained in section 2.3.1 and shown in Figure 2.9. This leads to the following failure criteria:

Face yield: $$F_L = \frac{\hat{\sigma}_f}{\frac{f_M L}{h_d h_f} + \frac{6}{4 h_f^2 \beta_k}}, \tag{2.32}$$

Core shear: $$F_L = \frac{\hat{\tau}_c h_d}{f_T}, \text{ and} \tag{2.33}$$

Core compression: $$F_L = \frac{2 \hat{\sigma}_c K_z h_c}{\beta_k E_f} = \frac{2\hat{\sigma}_c}{\beta_k}. \tag{2.34}$$

Here the simplified expression of K_z for $h_f \approx h_c$ from equation (2.5) is used. The resulting failure mode map shows that this assumption is valid for most relevant geometries, particularly for the transition from core shear failure to either face sheet fracture or core compression. The above equations are now generalized using the following expressions:

$$h^* = h_f/h_c, \quad L^* = L/h_c, \quad E^* = E_f/E_c,$$
$$\sigma^* = \hat{\sigma}_f/\hat{\sigma}_c \quad \text{and} \quad S^* = \hat{\sigma}_c/\hat{\tau}_c. \tag{2.35}$$

The transition lines between the different failure criteria can now be determined by equating the above failure criteria. Lim et al. [Lim04] showed this for the 3PB load case. The more general form now becomes:

(a) transition between face sheet fracture and core shear failure

$$1 = \frac{f_M L^*}{f_T s^* \sigma^* h^*} + \frac{3(1+h^*)}{2 f_T s^* \sigma^* h^*} \sqrt[4]{\frac{E^*}{3h^*}}, \tag{2.36}$$

(b) transition between face sheet fracture and core compression failure

$$1 = \frac{2\,f_M\,L^*}{\sigma^*(1+h^*)}\sqrt[4]{\frac{E^*}{3h^*}} + \frac{1}{\sigma^*}\sqrt{\frac{3E^*}{h^*}}\text{, and} \tag{2.37}$$

(c) transition between core shear failure and core compression failure

$$1 = \frac{2\,f_T\,s^*h^*}{(1+h^*)}\sqrt[4]{\frac{E^*}{3h^*}}\,. \tag{2.38}$$

For 3PB the load factors f_M and f_T now become [Zen97]:

$$f_M = 1/4 \quad \text{and} \qquad f_T = 1/2\,. \tag{2.39}$$

Applying equations (2.36) to (2.39) a failure mode map is created for a sandwich beam with a Rohacell 71RIST foam core and shown in Figure 2.14. The failure mode map follows the idea of Lim et al. [Lim04] and uses L^* and h^* on the horizontal and vertical axes. This aims at analyzing the effect of changing the geometric sandwich configuration instead of the core density as previously done by Triantafillou and Gibson [Tri87].

The results reveal that failure of short sandwich beams is driven primarily by the core behavior while increasing the span of the sandwich beam leads to a more face sheet driven failure behavior. Increasing the thickness of the face sheets relative to the core in turn causes the failure behavior of short sandwich beams to switch from core compressive failure, which is typically followed by local face sheet fracture, to core shear failure.

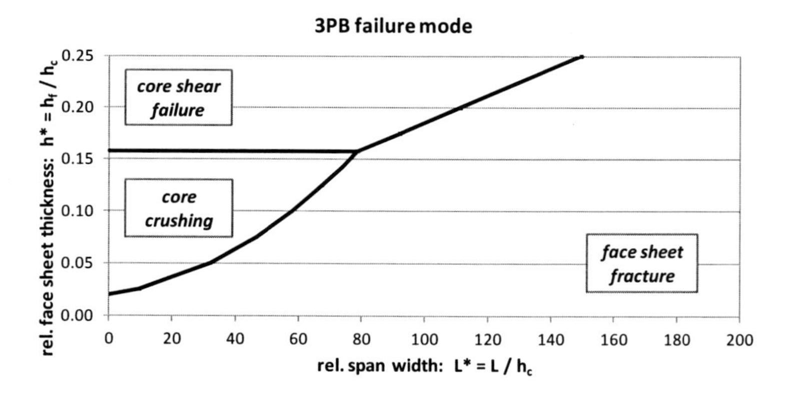

Figure 2.14 – three-point bending failure mode map of a sandwich beam with a Rohacell 71RIST foam core

In order to study the effect of different foam core densities Figure 2.15 shows failure mode maps of three different foam core densities. The selected densities are 51, 71 and 110RIST. Increasing the foam core density and thus strengthening the core leads generally to a more face sheet driven failure behavior.

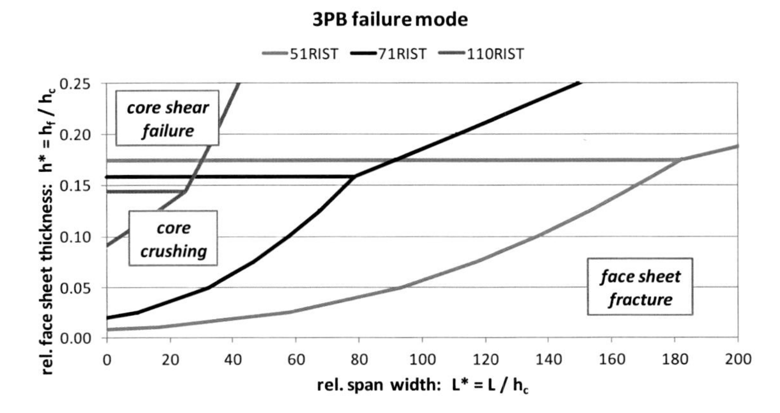

Figure 2.15 – three-point bending failure mode map of a sandwich beam with Rohacell 51, 71 and 110RIST foam cores.

2.4 Chapter summary

State of the art composite sandwich technology focuses in the aerospace industry on honeycomb cores due to their superior mechanical performance compared to foam cores. Honeycomb cores however suffer from some inherent drawbacks such as the sensitivity of the bond line, danger of water accumulation and high manufacturing costs. Industrial application of composite sandwich structures in other industries typically utilize polymeric foam cores due to significant cost benefits from simplified manufacturing and processing of large integral structures.

Currently research in the aeronautics focuses on exploitation of the cost benefits of polymeric foam cores for composite sandwich structures. Integrated core reinforcements such as pins or longitudinal profiles have the potential to improve the mechanical performance of foam cores significantly and thus make them more competitive to honeycomb cores.

The structural concept of a sandwich structure is very efficient by improving the bending stiffness and strength of large plane load bearding structures at only a modest weight increase when compared to monolithic structures. Related to this is an improved stability behavior. As a drawback sandwich structures are, due to the soft core, more sensitive to localized loads. These cause two kinds of loading, on the one side a local indentation and on the other side a global bending of the sandwich structure. This can be described by applying superposition within the limits of a material and geometric linear response.

Localized loads may occur due to load introductions or foreign object impacts. By applying a localized load on a sandwich beam the otherwise three dimensional impact problem on a sandwich plate is simplified to a two dimensional problem which helps understanding the underlying relationships.

The failure behavior of sandwich structures is more complex compared to monolithic composite structures. The different constituents of the structure may fail separately from one another leading to a larger degree of complexity. Failure modes relevant to the impact response are face sheet fracture, core shear failure and core crushing or local indentation.

A failure mode map can be created for simple structures that are subject to a defined load case. This map shows which kind of failure will occur depending on selected geometrical or design parameters and thus provides a design guideline to prevent a particular failure mode such as e.g. core shear failure in the case of a sandwich beam.

Failure mode maps are created for sandwich beams made of CFRP face sheets and a PMI foam core subject to three-point bending. Core shear failure can be prevented for most relevant span width of sandwich beams by choosing a sufficient core thickness. Increasing the core density and thus core strength improves this relation further.

In conclusion the failure behavior of short and thin sandwich beams is driven by the core properties while failure behavior of long and rather thick sandwich beams is instead driven by the face sheet properties.

3 Impact response of composite sandwich structures

3.1 Definitions

Damage tolerance is a design requirement of primary load bearing structures in safety critical applications such as aeronautics or large civil engineering structures. It describes "the ability (of a structure) to resist fracture from preexistent cracks (or defects) for a given period of time (and thus loads)" ([Gra03] p.3). Furthermore it "is an essential attribute of components whose failure could result in catastrophic loss of life or property" ([Gra03] p.3). In this context impact resistance can be described as the ability of a structure to resist the impact of a foreign object resulting in limited and manageable damage that does not immediately endanger the safe operation of a component or structure. Thus strictly speaking damage tolerance and impact resistance are separate design requirements but closely related as impact damage is a frequent source of defects that are accounted for during damage tolerance analysis. Therefore impact resistance is often considered as a sub topic of damage tolerance. The terms impact response and the more general impact behavior differ from the term impact resistance, as they describe the mechanical response and underlying damage mechanics during foreign object impact while impact resistance characterizes only its structural performance.

Primary load bearing applications in civil transport aircraft are in principal required to be designed damage tolerant. This typically implies that a structure contains at all times undetected defects, which must not impair the structure's ability to withstand all possible loads experienced during operation [Gra03]. If these defects grow under operating conditions and lead to a potentially catastrophic failure, they must be detectable via routine inspections prior to reaching a critical size. Besides failures dating back to the manufacturing process, a common source of structural defects are foreign object impacts which may occur during operation and maintenance. Composite materials typically have a more brittle impact response compared to most engineering metals and suffer from their low interlaminar shear strength. This applies similarly for composite sandwich structures.

Application of composite sandwich structures in the marine industry and in wind turbine blades show that this setback may be overcome by development of a thorough understanding of the underlying mechanics and related design methods. This can best be seen by the large number of publications such as e.g. [Zen97, Dan09] as well as products such as the Visby class corvette warship or the large number of wind turbine rotor blades. As the applied materials and design requirements however differ from those of the aerospace industry, a straight forward transfer of knowledge is difficult.

State of the art for certification of composite materials in primary aircraft structures to the regulatory requirements of the European Aviation Safety Agency (EASA) and Federal Aviation Authority (FAA) is known as "certification by analysis supported by test evidence" or "allowable-based certification" [Fer10]. It includes experiments used to identify material specific values for stiffness, strength and e.g. damage tolerance performance, as well as analysis methods applied for dimensioning and design. Analysis methods may be of both analytical and/or numerical nature. Testing and analysis are used in a mix in order to support and better understand each other's results. This process is often pictured by a pyramid and known as the building block approach (BBA), in which the most generic type of testing and analysis is placed at the base of the pyramid with stepwise more specific tests placed above leading to full scale testing at the top of the pyramid. "By combining testing and analysis, analytical predictions are verified by test, test plans are guided by analysis, and the cost of the overall effort is reduced while reliability is increased" [Mil02].

Application of the BBA to composite sandwich structures is an extensive, long-lasting program as a substantial number of aspects have to be addressed. Sandwich structures are particularly sensitive to impact and their response has been found to be very complex. Thus the BBA may be used to characterize the impact response of sandwich structures to minimize this specific risk and reach a satisfactory level of confidence. The generic BBA (Figure 3.1, left) starts with material coupons for determination of baseline material allowables, then rises to the element and detail level, where test specimens remain generic but incorporate specific characteristics of their later application. Components and subcomponents now become structural specific and form integral parts. On top of this is finally the full scale test for verification of design and analysis methodologies as required by certification. Applying this to the impact response of composite sandwich structures, the pyramid can be filled with tests on the coupon, element and detail levels, while the higher subcomponent, component and full scale levels are structure specific (Figure 3.1, right).

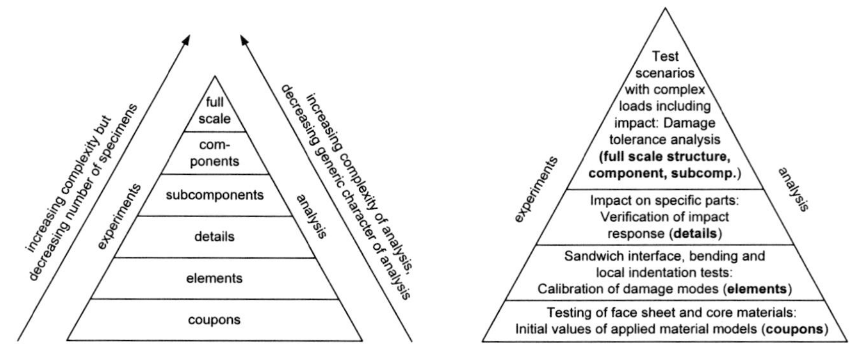

Figure 3.1 – building block approach (BBA): Generic (left) and applied to the impact response of composite sandwich structures (right)

Generalizing the impact response is difficult as different scenarios exist that differ significantly in size, type and shape of the impactor, impact velocity and energy. One common impact scenario is the tool drop. This scenario is a low velocity impact and describes a heavy tool such as e.g. a heavy wrench being dropped on the structure. It is relevant for most aircraft structure and may occur during manufacturing, operation and maintenance.

3.2 Literature review

An extensive review of past research on the impact response of composite sandwich structures is given by Abrate [Abr97, Abr98]. Abrate concentrates on the principal mechanics involved but also provides an overview of past publications sorted by face sheet and core materials. It is stated that most studies investigate only the impact response of a few sandwich configurations while varying the projectile parameters mass, shape and kinetic energy. In consequence the results of the reviewed studies are often in conflict with each other since the impact response of sandwich structures is influenced by parameters not necessarily covered by the individual studies. Abrate also provides an overview of investigations covering residual strength properties of sandwich structures.

Abrate differentiates between five different damage categories for honeycomb core sandwich structures affecting either the core or the impacted face sheet: (1) core buckling, (2) core cracking as well as (3) delaminations, (4) matrix cracking and (5) fiber cracking in the face sheet. The backward face sheet is typically not affected. Face sheet damage is similar to impact damage of monolithic laminates. Here damage size is driven by delaminations and increases almost linearly with impact energy until it becomes visible. A further increase of the impact energy beyond the visibility threshold does not anymore increase the affected area significantly. Core damage consists of buckling or cracking of cell walls in a region surrounding the impact point which is typically referred to as core crushing.

Bernard et al. [Ber89] compare in their study the impact behavior of honeycomb and foam cores. The behavior of the foam differs to honeycomb in that at low impact energies the core damage appears like a vertical crack while at higher energies the damage looks more like a depression below the point of impact. The size of the core damage increases with impact energy similarly as observed for the face sheets.

Finally Abrate refers to two studies that aim specifically into finding governing parameters that affect the failure behavior of composite sandwich structures. The first mentioned is from Triantafillou and Gibson [Tri87] and describes static three-point bending (3PB) tests of sandwich beams made of aluminum face sheets and polyurethane foam core. Triantafillou varies the loading within the sandwich beam by adjusting face sheet and core thicknesses, foam core density and the distance between the supports in the 3PB test. Triantafillou then derives an analytical approach for describing the three main governing failure modes of face sheet yield, face sheet wrinkling and core shear failure. Agreement of experimental and analytical results is reported well and summarized in damage mode maps which visualize the results graphically. Depending on either of the varied parameters different failure modes are triggered as already discussed in section 2.3.2.

The use of static 3PB tests for understanding the impact behavior of composite sandwich structures might appear surprising. The 3PB test can however be understood as the two dimensional, static simplification of the otherwise three dimensional, dynamic sandwich impact problem as suggested and applied by Shipsha et al. [Shi03, Shi03a]. Similarly

Petras and Sutcliffe [Pet00] use 3PB tests of sandwich beams to demonstrate the effect of combined compression and shear loading in Nomex honeycomb cores on sandwich failure as shown in Figure 3.2.

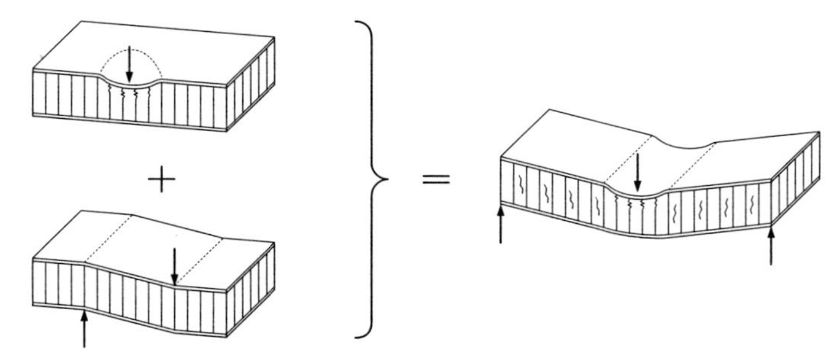

Figure 3.2 – schematic of honeycomb core sandwich failure mechanisms [Pet00]

The second study mentioned by Abrate is from Mines et al. [Min94]. It also describes 3PB tests of sandwich beams covering composite face sheets made of glass, aramid or carbon fibers with epoxy resin that are bonded to a honeycomb core. Loading is applied both statically and dynamically. Mines identifies four failure modes: (I) Upper skin compressive failure followed by (a) core crushing or (b) core shear failure with the later absorbing significantly less energy. The face sheets of affected specimens have a lower compressive than tensile strength. (II) Compressive failure of the upper skin proceeded by a tensile failure of the lower skin typically occurs as woven aramid skins are used. The upper skin fails first due to compression but remains intact until final fracture occurs. (III) Core shear failure is related to an initial failure of the interface of the upper skin to core. (IV) Tensile failure of the lower skin without any preceding damages may occur when glass continuous strand mats are used for the face sheets as these have – opposite to the other face sheet materials – a lower tensile than compressive strength.

Lim et al. [Lim04] perform static and dynamic 3PB tests of sandwich beams. The beams were made of GFRP face sheets and PVC foam core. Based on an analytical approach already described by Zenkert [Zen97] and Thomsen [Tho92], a damage mode map with variable geometric parameters is presented. Lim et al. distinguish between three major damage modes of the sandwich beam: (1) Face sheet fracture, (2) core shear and (3) core compressive failure. Experimental results of the static tests agree well. Dynamic tests are instead correlated with explicit FEM calculations.

A major survey of the damage tolerance of composite sandwich airframe structures was started by the Federal Aviation Administration (FAA) in the late 1990's which resulted in several publications and technical reports. The performed work focuses on honeycomb core structures with CFRP face sheets as applied in general aviation aircraft and business jets. Tomblin, Raju et al. of Wichita State University conduct both impact and residual strength tests on sandwich structures in accordance to the building block approach. Despite using a honeycomb core material much can be learned regarding the mechanics of sandwich structures during impact loading and related testing methods.

At first Tomblin et al. [Tom99] provide a review of previous damage tolerance investigations including some reference sandwich configurations of aircraft currently in service as well as abstracts of discussions with different manufacturers regarding their design procedures. The subject of impact resistance is usually accounted for by experimentally determined allowable damage limits and critical damage thresholds. In parallel Moody et al. [Moo00] present in a companion study the current status of analytical methods for assessing the damage tolerance of composite sandwich structures.

Following this Tomblin et al. [Tom01, Tom02] perform impact and compressive residual strength testing of CFRP honeycomb sandwich structures with varying sandwich configurations. They conclude that dent depth and thus damage visibility does not correlate well with residual strength as core damage can be hidden by spring back of the face sheet while damage area correlates significantly better. Damage area increases with impactor diameter and impact energy but decreases with core thickness. Increasing the face sheet thickness may also be detrimental as published later by Raju and Tomblin [Raj08]. Based on these results Lacy et al. [Lac02, Lac02a] develop response surfaces for impact resistance and residual strength which reproduce the basic findings of the experiments.

Finally Tomblin et al. [Tom04] investigate scaling effects and conclude that the impact response does not alter in principal due to scaling of the in-plane sandwich dimensions. As larger sandwich plates however store more energy elastically, the damage size related to specific impact energies is larger on small specimens than on large specimens. Placing impacts off center closer to the panel edge has the same effect as it reduces the effective panel size. The compression after impact (CAI) strength is driven by damage area increasing slightly with specimen size. Transfer of results from tests on smaller specimens to larger specimens based on damage size is thus conservative. Tomblin et al. [Tom05] also investigate different face sheet layups but report only minor effects. The use of an open hole in the face sheet as an equivalent impact damage during CAI testing is considered as a worst case representation.

Edgren et al. [Edg04, Edg08] investigate the failure processes of impacted composite foam core sandwich panels during CAI. The composite face sheet was made of CFRP using non-crimp fabrics (NCF) and resin infusion. Less attention is put on the impact response as only two reference impact energies are applied that correlate with the barely visible impact damage (BVID) and visible impact damage (VID) thresholds. Characterization of the failure process reveals that specific for NCF composites specimen failure is preceded by significant local fiber buckling leading to the formation of kink bands. Equivalent open hole representations of the impact damage in the impacted face sheet are developed and compared with a notch representation. The notch representation is recommended while the results of the open hole representation differ only by 5 to 10%.

Leijten et al. [Lei09] present impact test results performed on a similar material combination as used within this work. Baseline is a Rohacell 110XT-HT PMI foam core with quasi-isotropic CFRP face sheets. Boundary conditions as suggested by Airbus [Air05] were applied. Leijten et al. [Lei09] conclude that the impact response can be characterized by three distinct regions of impact energy as shown in Figure 3.3. In the first region damage size increases linearly with impact energy, the second region describes a plateau with constant damage size while the third region describes perforation of the impacted face sheet.

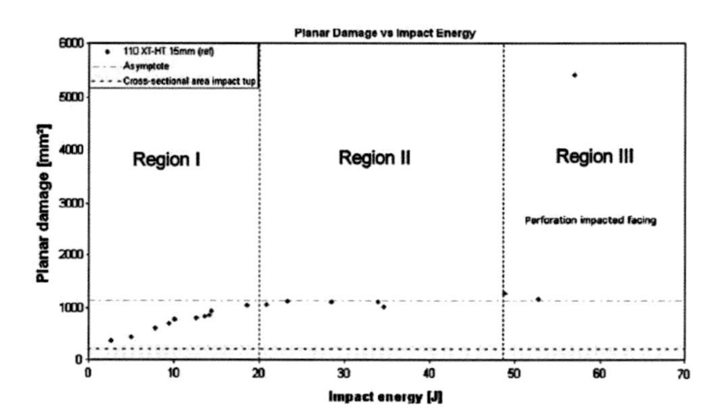

Figure 3.3 – damage area vs. impact energy with regions [Lei09]

Leijten et al. [Lei09] analyzed different foam core densities and Rohacell PMI foam types with respect to their BVID threshold describing noticeable effects on damage size and visibility while the principal behavior remains identical. CAI results are presented as a measure of the damage tolerance capability. Furthermore Leijten et al. [Lei09] describe that a change of local to global damage behavior takes place when using significantly smaller core thickness than the reference configuration. It is stated that this change is related to the smaller sandwich bending stiffness which leads to core shear failure due to impact.

Composite honeycomb core sandwich structures were also investigated by Meo et al. [Meo05], Heimbs [Hei08], Kärger et al. [Kae08] and Manes et al. [Man13]. Sandwich structures with folded cores were subject of investigations by Nguyen et al. [Ngu05], Heimbs et al. [Hei10] and Klaus et al. [Kla12] while foam core sandwich structures were investigated by Schubel et al. [Sch05], Wang et al. [Wan13] and Möhle et al. [Moe12].

Analytical models of the impact process can be used for classification of the response. Chai and Zhu [Cha11] describe in their review analytical models for low velocity impact on sandwich plates. In a separate work Rahammer [Rah11] performed a review of quasi-static indentation models with focus on their capability to predict delaminations in composite sandwich panels with a honeycomb core. Analytical models that describe the indentation and impact process were also published by Olsson [Ols02] and Zhou [Zho06].

3.3 Classification of the structural impact response

Impact damage in composite sandwich structures – commonly referred to as foreign object damage – can be very different depending on velocity, mass, type and shape of the impacting object as well as size, type and boundary conditions of the structure being hit. Analytical models may be used for classification of the impact response. Generally impact on a sandwich structure causes two types of responses which shall be treated separately in order to allow a better understanding of the governing factors.

As the impactor hits a sandwich structure, both objects get into contact which causes locally a contact indentation problem which dominates the local response. On top of this the sandwich structure itself is – due to the pressure distribution generated by the local indentation problem – also loaded globally. This global response depends on the size of the plate and its boundary conditions as well as mass and velocity of the impactor and the sandwich itself. Both the local and global responses are often superposed although nonlinearities such as e.g. local material failure occur and therefore prohibit linear superposition. By applying superposition however the complete response of the plate can be described as shown in Figure 3.4. This approach is thus limited to small nonlinearities.

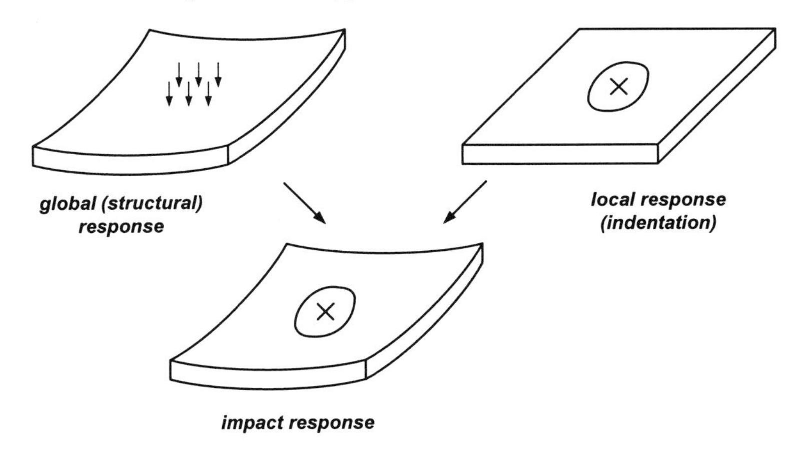

Figure 3.4 – mechanical response of a sandwich plate subject to impact loading; global (structural) and local (indentation) responses add to the impact response

Often the impact energy is used as reference value to characterize the severity of an impact. This works reasonably well for impacts of similar conditions e.g. the same type of structure being hit by the same impactor but with a similar velocity or mass. If however the velocity or mass change significantly, this type of comparison is not anymore true, as inertia effects in the structure being hit increase significantly as the impact velocity rises. One common mistake is that the change of the impact response is related to strain rate dependent behavior of the material. It is well known that materials act differently when subject to varying strain rates. Strain rate dependent behavior however adds to the change of the structural response, but is not necessarily the reason of the later.

The impact response of a plate can be classified into different categories. Typically three response types – low velocity, high velocity and ballistic impact – are distinguished. This implies at first sight that the velocity of the impactor determines the type of impact. The velocity is an important parameter, neglects however the characteristics of the plate being hit as well as mass and thus inertia effects of both impact partners as well as the material properties. Thus a more refined approach becomes necessary.

In his publication dedicated to impact on sandwich structures Abrate [Abr97] does not discuss the effect of velocity in detail but proposes in his more general work [Abr98] a simple way to determine a transitional velocity between ballistic and non-ballistic impacts on monolithic laminates which provides already some guidance. Once the impactor gets into contact with the plate, compressive stress waves, shear waves and Rayleigh waves propagate outward from the point of impact. Compressive and shear waves reach the back face, reflect backwards and establish the plate motion. If the plate motion is established before the impactor causes significant damage, Abrate [Abr97] calls this a low velocity impact as the plate behavior of the target being hit dominates the impact response. Otherwise these impacts are called ballistic impacts as their behavior is dominated by stress wave propagation in the thickness direction of the plate.

For a monolithic composite the transitional velocity between plate motion and through the thickness stress wave dominated behavior can be calculated by assuming a cylindrical rod being hit by a rigid mass. The rod represents the circular area of the plate being hit. The initial compressive strain ε of the impacted surface can now be described by [Abr98]

$$\varepsilon = v_0 / c \,. \tag{3.1}$$

Here v_0 is the initial velocity of the impactor while c is the speed of sound in the thickness direction of the material. If the initial compressive strain ε now surpasses the critical strain level for the impacted material, stress waves in the thickness will influence the structural behavior significantly. For composite materials with an epoxy matrix Abrate [Abr97] reports a critical strain level on the order of 1%. Based on a speed of sound of 2000 m/s [McH01] for RTM-6 resin, which governs the material properties in the thickness direction of the composite laminate, the transitional velocity is on the order of 20 m/s.

Olsson [Ols00] uses a more detailed approach for composite plates and distinguishes between low velocity, high velocity and ballistic impact as three principal types (see also Figure 3.5). For non-ballistic impacts also stress wave propagation in the longitudinal direction of the plate is considered. Impacts that can be described using the quasi-static response of the plate are low velocity impacts, while those that are dominated by in-plane stress wave propagation are called high velocity impacts. Non-ballistic impacts that share the characteristics of both low and high velocity impacts thus fall into neither category and are referred to as intermediate velocity impacts.

As Olsson [Ols02] applies in his work a mass criterion to distinguish between low, intermediate and high velocity impacts, he also refers to large mass (low velocity), medium mass (intermediate velocity) and small mass (high velocity) impacts. For sandwich structures the same mass criterion has been applied [Ols02].

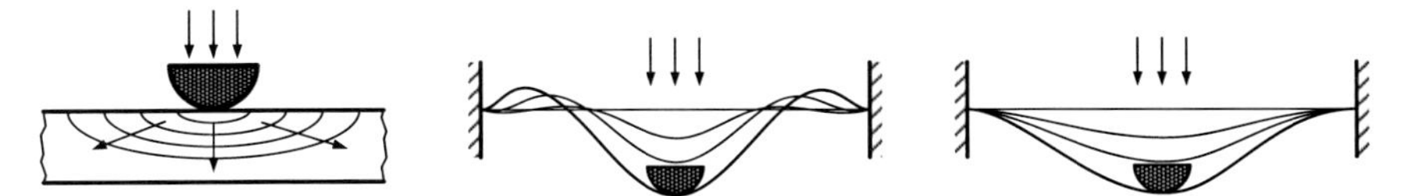

Figure 3.5 – response types of impact on composite plates [Ols00]:
Ballistic impact (left), high velocity impact (center) and low velocity impact (right)

A similar distinguishing has been done by Fatt et al. [Fat10]. Here a model for the high velocity impact response was developed. A model for the low velocity response was presented in previous work [Fat01, Fat01a]. Chai and Zhu [Cha11] propose in their extensive review a stepwise approach to categorize impact on sandwich plates. Figure 3.6 shows the basic model used and the related properties of the sandwich structure.

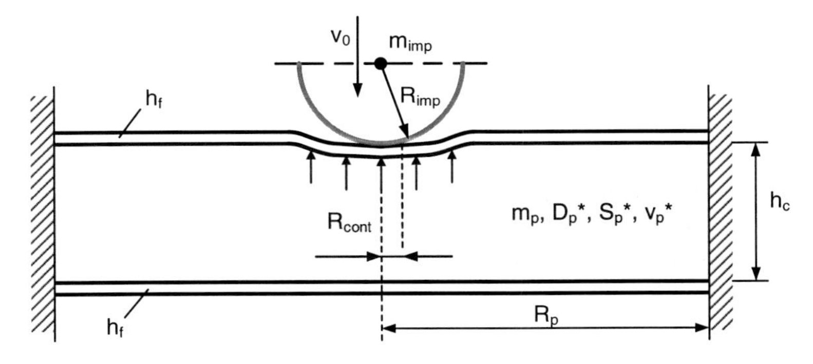

Figure 3.6 – basic model for impact on a sandwich panel [Cha11]

At first impacts are categorized into ballistic and non-ballistic behavior depending on the contact time t_{cont}. If the contact time is on the order of magnitude of the time required by the stress waves for propagation through the thickness of the impacted material, the impact is of ballistic behavior. With h designating the specimen thickness, this ballistic contact time $t_{cont,bal}$ can be expressed by

$$t_{cont,bal} = \frac{h}{\sqrt{E/\rho}} \,. \tag{3.2}$$

Here the square root of E/ρ approximates the speed of sound c in the impacted material. For thin monolithic composite plates similar to face sheets typically used in sandwich structures, the time for stress wave propagation can be calculated using equation (3.2). For

CFRP of 1.5 mm thickness, a density of 1550 kg/m³ and a stiffness of 10000 MPa in the thickness direction stress waves t_{face} becomes

$$t_{face} = \frac{1{,}5\text{mm}}{\sqrt{10\frac{\text{kg}}{\text{mm}*\text{ms}^2}/1550*10^{-9}\frac{\text{kg}}{\text{mm}^3}}} \approx 0.0005905 \text{ ms}.$$

This can now be used to judge whether the response of a monolithic plate is controlled by stress wave propagation in the thickness direction. For a contact time t_{cont} that is in the same order of magnitude as t_{face}, the response is ballistic.

This idea can also be extended to the foam core. For a Rohacell 71RIST foam core with 25.4 mm thickness, a stiffness of 105 MPa and a density of 75 kg/m³ the time for stress wave propagation on the thickness direction of the foam core is

$$t_{core} = \frac{25{,}4\text{mm}}{\sqrt{0{,}105\frac{\text{kg}}{\text{mm}*\text{ms}^2}/75*10^{-9}\frac{\text{kg}}{\text{mm}^3}}} \approx 0.02147 \text{ ms}.$$

One has to keep in mind that this describes the stress propagation through the sandwich core only. Nevertheless it becomes clearly visible that the response of the core is much more sensitive to ballistic behavior than the response of the face sheets. Thus the sandwich response is in this case governed by the core behavior.

Applying the same criteria to the stress wave propagation in the sandwich plane, non-ballistic impacts can be divided into quasi-static behavior (low velocity) and stress wave dominated behavior (high velocity). In this case the material properties are of concern in the direction of the wave propagation. The plate radius R_p, which is equivalent to the smallest distance from the impact point to the closest boundary condition and shown in Figure 3.6, now becomes the relevant length to be traveled by the stress waves. The contact time of a high velocity impact $t_{cont,wave}$ now becomes

$$t_{cont,wave} = \frac{R_p}{\sqrt{E/\rho}}. \tag{3.3}$$

As an example the critical contact duration is calculated for a sandwich structure made of the CFRP and PMI foam used in the previous examples but with an in-plane CFRP stiffness of 50000 MPa typical for a quasi-isotropic laminate. For the distance $R_p = 100$ mm has been chosen:

$$t_{in-plane} = \frac{100\text{mm}}{\sqrt{50\frac{\text{kg}}{\text{mm}*\text{ms}^2}/1550*10^{-9}\frac{\text{kg}}{\text{mm}^3}}} \approx 0.01761 \text{ ms}.$$

Comparing the critical contact times it becomes clear that the impact behavior of the face sheet itself is much less suspect to stress wave dependent behavior than the behavior of the core or the in-plane behavior of the sandwich structure. For the two later mentioned cases the critical contact time is of the same order of magnitude. Thus for a sandwich panel with only 100 mm distance from the impact point to the supported edges, stress wave propagation takes about the same amount of time in-plane and through the thickness.

This is different but not necessarily in contrast to what is commonly used. In the literature the mass ratio $\bar{m}$ of the impactor mass m_{imp} to the effective mass of the plate m_{ef} is commonly used as criterion for differentiating between high and low velocity impacts [Ols00, Zho06, Zho07, Cha11]. The mass ratio $\bar{m}$ is now defined as:

$$\bar{m} = \frac{m_{imp}}{m_{ef}} . \quad (3.4)$$

The effective mass m_{ef} of the impacted plate depends on the energy balance and the stiffnesses of the sandwich plate, but is typically less than or equal to ¼ of the total mass of the plate m_p [Zho06]. Chai and Zhu [Cha11] point out that the criteria for low velocity impacts of Olsson [Ols00], Zhou and Stronge [Zho06] and Swanson [Swa72] can be unified. This is based on Swanson's finding that quasi-static solutions for impacts on plates are only true if the ratio of the effective mass $\bar{m} > 8$ where $\bar{m}$ is calculated using the following rule of thumb:

$$\omega_i < \frac{1}{3}\omega_p . \quad (3.5)$$

Here ω_i is the impact frequency and ω_p the fundamental frequency of the plate. Olsson [Ols00] uses for a simplified mass criterion instead the total mass of the impacted plate. Considering that $m_{ef} \leq$ ¼ m_p Olsson's simplification is valid with respect to equation (3.5).

In summary non-ballistic impacts can be classified using the following mass ratios. These have been developed for use with monolithic composite plates [Ols00] but also applied successfully to sandwich plates [Ols02]:

$$\frac{m_{imp}}{m_p} > 2 \qquad \text{(low velocity / large mass impact),} \quad (3.6)$$

$$2 > \frac{m_{imp}}{m_p} > \frac{1}{5} \qquad \text{(intermediate velocity / medium mass impact),} \quad (3.7)$$

$$\frac{1}{5} > \frac{m_{imp}}{m_p} \qquad \text{(high velocity / small mass impact).} \quad (3.8)$$

The stepwise approach to characterize the structural response of composite plates summarized from the literature by Chai and Zhu [Cha11] is also shown in Figure 3.7.

Different models have been proposed to describe the impact response of sandwich panels. Rahammer [Rah11] provides in his work an overview of available sandwich indentation models but neglects the structural response. Chai and Zhu [Cha11] discuss both the indentation and the structural response separately and thus provide a good overview.

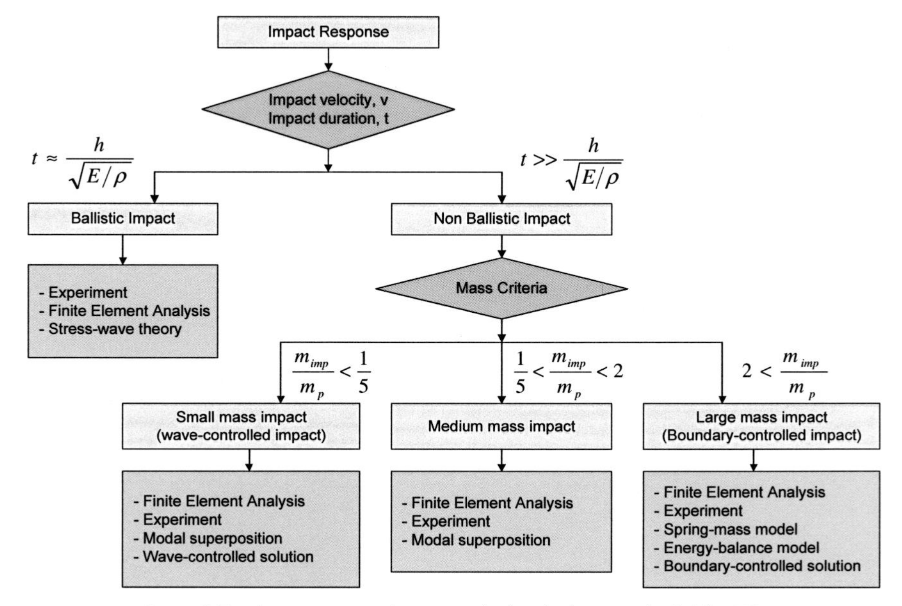

Figure 3.7 – characterization of impact and related solution methods [Cha11]

In principal one can distinguish between spring-mass models, energy balance models and modal superposition methods as already shown in Figure 3.7. Alternatively a spring mass damper model has been proposed by Olsson and Davies [Ols02, Dav04, Ols10]. Spring mass and energy balance models are only applicable to low velocity impact while the modal superposition method and the spring mass damper model are applicable to intermediate and high velocity impacts.

Spring mass models use a dynamic model of the impactor mass $\mathrm{m_{imp}}$ that is slowed down by a set of springs representing the local contact stiffness in case of the impactor and the combined global stiffnesses of the sandwich in case of the sandwich plate. More sophisticated models are based on a two mass spring system based on the impactor mass $\mathrm{m_{imp}}$ and the effective dynamic mass of the impacted plate $\mathrm{m_p^*}$ as presented in Figure 3.8 (left) based on work of Olsson [Ols02].

Here the mass of the impactor is slowed down by a single spring which represents the contact and indentation stiffness $\mathrm{k_h}$ of the sandwich plate. This spring decelerates the impactor but accelerates the mass of the sandwich plate which itself is placed on a set of two parallel springs. The first of these springs is the combined bending stiffness $\mathrm{k_b}$ and shear stiffness $\mathrm{k_s}$ which slows down the mass of the plate relative to the global deflection of the plate. The second parallel spring is the membrane stiffness $\mathrm{k_m}$ which becomes effective for larger global plate deflections.

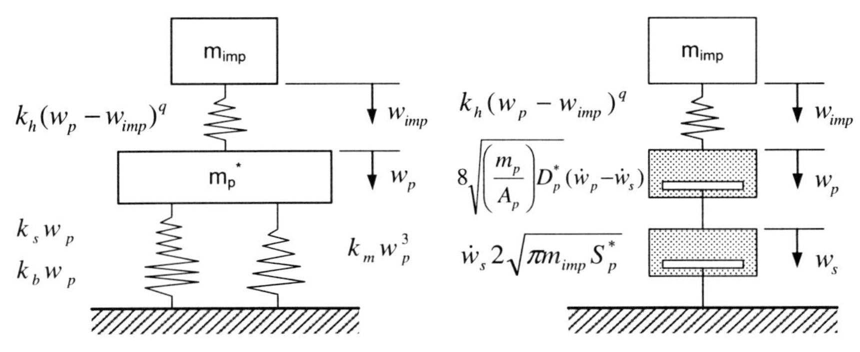

Figure 3.8 – sandwich impact models: Spring mass model for low velocity sandwich impact (left) [Ols02], spring mass damper model for intermediate and high velocity sandwich impact (right) [Dav04]

The spring mass model is applicable only to large mass impact as characterized in equation (3.6). Thus quasi-static loading conditions can be assumed for impactor masses m_{imp} that are more than twice as large as the plate mass m_p. This leads to the conclusion that inertia effects of the impacted sandwich plate can be neglected. Equilibrium of forces must be present between the contact indentation force represented by k_h and the plate reaction force represented by the combined plate bending and shear stiffness k_{bs} and the membrane stiffness k_m [Ols02]. This can be expressed by

$$F_{imp} = k_h\left(w_p - w_{imp}\right)^q = k_{bs} w_p + k_m w_p^3 \quad \text{with} \quad \frac{1}{k_{bs}} = \frac{1}{k_b} + \frac{1}{k_s}\,. \tag{3.9}$$

The bending, shear and membrane stiffnesses of the sandwich plate can be calculated using sandwich plate theory as explained in appendix A.2. Solutions for a point load on a sandwich plate can be found in common literature on sandwich structures such as e.g. Zenkert [Zen95]. The indentation work W_γ that is consumed by the sandwich through local indentation can be described by [Ols02]

$$W_\gamma = \int F_{imp} d\gamma = \int k_h \gamma^q d\gamma = \frac{k_h \gamma^{(q+1)}}{q+1} = \frac{F_{imp}^{\left(1+\frac{1}{q}\right)}}{(q+1) k_h^{1/q}}\,. \tag{3.10}$$

Here q is a constant that describes the contact relation. It has to be kept in mind that this equation may only be valid as long as no damage occurs in the sandwich as otherwise k_h will change. Thus in reality it is often necessary to solve this equation stepwise and adjust k_h with growing damage. Combining equations (3.9) and (3.10) the work of the impactor W_{imp}, equivalent to the applied impact energy E_{imp}, can be described by [Ols02]

$$W_{imp} = \frac{1}{2}k_{bs}w_0^2 + \frac{1}{4}k_m w_0^4 + \frac{(k_{bs}w_p + k_m w_p^3)^{\left(1+\frac{1}{q}\right)}}{(q+1)k_h^{1/q}} \,. \tag{3.11}$$

This leads to energy balance models which utilize the principle of conservation of total energy within the system of the impactor and the structure. Here the focus is put on the energy consumption of an indenter as it is pushed into the sandwich face sheet. The energy consumption includes both elastic deformation of the sandwich and the energy consumed as damage occurs during the indentation process. The approach of Zhou and Stronge [Zho06] is based on this and states that in the case of a heavy projectile hitting a light weight panel at low velocity, the kinetic energy of the impactor E_{kin} will be stored by elastic and plastic deformation of the panel as summarized in [Zho06]

$$E_{kin} = \frac{1}{2}m_{imp}v_0^2 = E_{bs} + E_\gamma + E_m \,. \tag{3.12}$$

Here E_{bs} is the energy stored by elastic bending and shear strains of the sandwich plate, E_m the energy stored due to membrane strains respectively while E_γ is the indentation energy. The individual energies can be described by [Zho06]

$$\begin{aligned} &E_{bs} = \frac{1}{2}k_{bs}w_0^2 \,, \qquad E_m = \frac{1}{4}k_m w_0^4 \qquad \text{and} \\ &E_\gamma = \chi_1 \int_0^{\gamma_{0m}} \sqrt{\gamma_0 + \chi_2\,\gamma_0^3}\, d\gamma_0 \,. \end{aligned} \tag{3.13}$$

Here γ_0 is the indentation at the point of impact and γ_{0m} the maximum indentation during the impact. The parameters χ_1 and χ_2 contain both geometric and material information. They are derived by Zhou and Stronge [Zho06] by minimizing the total potential energy of the indentation problem alone. Here core crushing as well as face sheet bending and local membrane stretching are considered while delaminations are ignored. For a Poisson ratio $\nu = 0.3$ they become

$$\chi_1 = \left(\frac{16\,\pi}{3}\right)\sqrt{D_f\, p_{cr}} \qquad \text{and} \qquad \chi_2 = \frac{0.488}{h_f^2} \,. \tag{3.14}$$

In quasi-static conditions there must be equilibrium of forces at the sandwich plate mass m_{pl} at a given deflection w_0 with the contact indentation force of the impacted face sheet on the one side and the reactive force of the sandwich plate at the other side. This can be used for the determination of γ_{0m} as expressed by Zhou and Stronge [Zho06]

$$k_{bs}w_0 + k_m w_0^3 = \chi_1\sqrt{\gamma_{0m} + \chi_2\,\gamma_{0m}^3} \,. \tag{3.15}$$

Zhou and Stronge write that in practice an initial contact radius R_{cont} is assumed for the determination of γ_{0m}. Equations (3.13) and (3.15) can now be solved using a numerical integration scheme such as e.g. the Newton-Raphson method. When comparing with Figure 3.8 (left), the left side of equation (3.15) is based on the reactive force of the plate represented by the two parallel springs of membrane and combined bending shear stiffnesses. The right side of the equation is equivalent to the indentation stiffness.

Laminated plate theory such as CLT or higher order approaches such as Mindlin-Reissner plate theory can be applied and used for calculation of the actual stress and strain states of the indented face sheet and its mechanical response. The plate is placed on the foam which is modeled as an initially elastic foundation and later in the indentation process ideally plastic foundation. As the out-of-plane deformations of the face sheet become larger, plate theory however becomes less accurate and has to be corrected for membrane stresses. This becomes particularly important as delamination damage occurs in the face sheet which eliminates its bending stiffness almost completely.

In summary both large mass impact models of Olsson [Ols02] and Zhou and Stronge [Zho06] can be unified as they both apply a spring mass system to describe the force vs. indentation relation and an energy balance method to describe the amount of energy consumed up to a certain deflection of the impactor.

The modal superposition method can be applied for intermediate and high velocity impacts and treats the system as a continuum with infinite degrees of freedom. For this Zhou and Stronge [Zho06a, Zho06] apply Mindlin-Reissner plate theory but neglect inertia effects and membrane stretching. For small masses impacting a relatively heavy structure the method captures the impact response well as different vibration modes are excited and can be calculated using this method. The experimental results are however limited to small damages and thus only small impact forces well below one kN. Thus it is not immediately clear if this method can be transferred to small mass impacts with greater impact energies that cause significant structural damage. As the proposed method of Zhou and Stronge neglects the effects of membrane stretching and inertia, it is not applicable to low velocity impact as these effects become more important here.

Finally Olsson and Davies [Ols02, Dav04, Ols10] developed a spring mass damper model as shown in Figure 3.8 (right). This model uses for the description of the local indentation the same approach as the low velocity impact model shown in Figure 3.8 (left). The structural response of the sandwich plate is however modeled depending on the change of the bending deflection $\dot{w}_p$ and the shear deformation $\dot{w}_s$. Thus the mechanical response is now dependent on the rate of change of the displacement and not anymore quasi-static. It is suitable for intermediate and high velocities but cannot be applied for ballistic impact.

3.4 Contact behavior and indentation of sandwich panels

A well-known theory for contact analysis of isotropic bodies is the Hertz law. It however does not compensate for the more complex setup of a composite sandwich structure with a low modulus core and composite face sheet and thus has to be modified [Cha11]. Additionally one has to keep in mind that as damage due to the contact accumulates in the face sheet and the core, the contact behavior also changes.

The basic contact situation of isotropic bodies subject to solely elastic deformation as described by Hertz law is shown in Figure 3.9. The contact pressure p can be expressed as

$$p = p_0\sqrt{1-\left(\frac{r}{R_{cont}}\right)^2}\ . \tag{3.16}$$

Here p_0 is the maximum contact pressure at the centre of the area of contact. R_{cont} and r are the radius of the contact zone and the radial position in the contact zone respectively. The force indentation law can now be expressed as

$$P = k_h \gamma^q \tag{3.17}$$

with k_h being the contact stiffness, γ impactor indentation and q a constant. Depending on the individual material properties different expressions for k_h and q have to be used and may need to be determined based on experimental results.

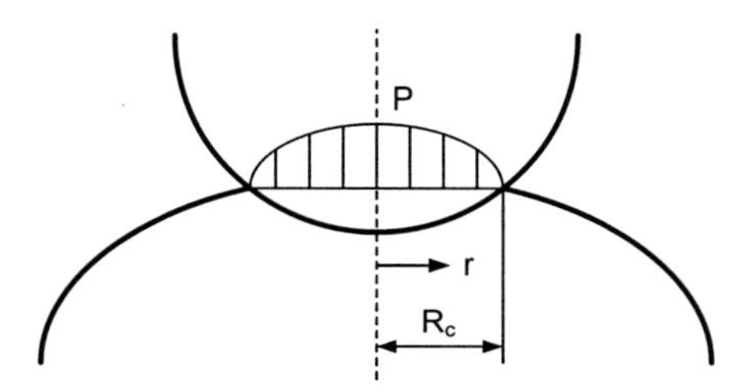

Figure 3.9 – Hertzian pressure distribution [Cha11]

For composite laminates the contact can be described using a formulation by Yang and Sun [Yan82] which is based on the two main influencing factors, the radii R_1 and R_2 of the two bodies in contact and their respective elastic moduli E_1 and E_2. The contact stiffness k_h now becomes a function of these two:

$$k_h = \frac{4}{3}E^*\sqrt{R^*}$$
$$\text{with} \quad \frac{1}{R^*} = \frac{1}{R_1}+\frac{1}{R_2} \qquad \text{and} \quad \frac{1}{E^*} = \frac{1-\nu_1}{E_1}+\frac{1-\nu_2}{E_2}\ . \tag{3.18}$$

Improved contact formulations for composite sandwich structures have been developed by Olsson [Ols02] and Zhou and Stronge [Zho06]. Here the radius of the contact zone R_c is described by:

$$R_{cont} = \sqrt{R_{imp}\,\gamma}\ . \tag{3.19}$$

The profile of the local indentation can then be expressed as [Zho06]

$$\gamma(r) = \gamma_0\left(1-\frac{r^2}{a^2}\right)\ . \tag{3.20}$$

Here $\gamma(r)$ is the local indentation with r being the distance from the center of the contact, γ_0 the maximum indentation and a the radius of the contact region. These equations are however only valid for quasi-static and thus low velocity impacts [Cha11]. Also these formulas have to be improved as damage such as core yielding or delaminations in the face sheet occur. Thus the Hertz law as described in equation (3.20) is only valid during the first elastic stage of the impact response which is dominated by the face sheet behavior.

Olsson investigated the indentation response in more detail [Ols94] and concluded that the parameter q from equation (3.17) varies depending on whether a monolithic composite plate or a sandwich structure is analyzed. For a composite plate $q = 1.5$ is recommended while for a sandwich structure instead $q = 1.0$ should be used. In case of unknown material combinations an experimental verification of this parameter should be performed as proposed by Chai and Zhu [Cha11].

The indentation of a sandwich panel describes only the local behavior of the sandwich as if it was placed on fixed boundary conditions. This is in contrast to the complete impact response which is the superposition of the nonlinear local indentation and global structural responses of the sandwich plate. Superposition may only be conducted as long as nonlinear behavior remains small.

The displacement of the impactor w_{imp} can now be split into different components, with w_p as the global displacement of the sandwich plate typically measured at the back side of the sandwich and the sandwich indentation γ. Sandwich indentation arises from the local response which includes Hertzian indentation of the face sheet γ_H and the impactor itself γ_{imp} as well as face sheet deflection due to core deformation γ_f. As Hertzian indentations are typically small, it is sufficient to concentrate on the face sheet deflection γ_f and neglect the Hertzian indentation for description of the local response. Thus the displacement of the impactor w_{imp} can be calculated using superposition of the local and global behaviors

$$w_{imp} = \gamma_f + w_p \; . \tag{3.21}$$

A detailed three step indentation model has been developed by Olsson [Ols02] and is schematically shown in Figure 3.10. This model is used as reference as it provides a good understanding of the damage processes that take place during the indentation.

In phase (1) the indentation is fully elastic, only elastic indentation occurs and the face sheet is modeled using plate theory. The plate itself is placed on the core which is modeled as a continuous elastic foundation as described in section 2.3.1. The indentation of the sandwich face sheet γ_f can now be expressed by [Ols02]

$$\gamma_f = \frac{\frac{1}{8}F}{\sqrt{K_z D_f^*}} \; . \tag{3.22}$$

Here K_z is elastic foundation stiffness from equation (2.5) and D_f^* is the effective plate bending stiffness of the face sheet. The calculation of effective plate and membrane properties for orthotropic face sheets and sandwich plates is found in appendix A. The elastic response of phase (1) ends as the core strength in the thickness direction is superseded and core crushing is initiated. This critical deflection γ_{cr} can be described by equation (3.22) and the critical load F_{cr} that initiates core crushing by [Ols02]

$$F_{cr} = 3\sqrt{3}\, p_{cr} \left(\frac{2\, D_f^*}{E_{zc}^*}\right)^{2/3} \quad \text{for} \quad h_f \ll h_c \quad \text{and}$$
$$F_{cr} = 8\, p_{cr} \sqrt{\frac{D_f^* h_c}{1.38\, E_{zc}^*}} \quad \text{for} \quad h_f \approx h_c\,. \tag{3.23}$$

Here p_{cr} is the compressive yield stress of the sandwich core and E_{zc}^* the effective stiffness of the sandwich core in the through the thickness direction. For most core materials p_{cr} is identical with the continuous crushing stress. The effective core stiffness in the thickness direction can be assumed for isotropic foam core materials by

$$E_{zc}^* = \frac{E_c}{1-\nu^2}\,. \tag{3.24}$$

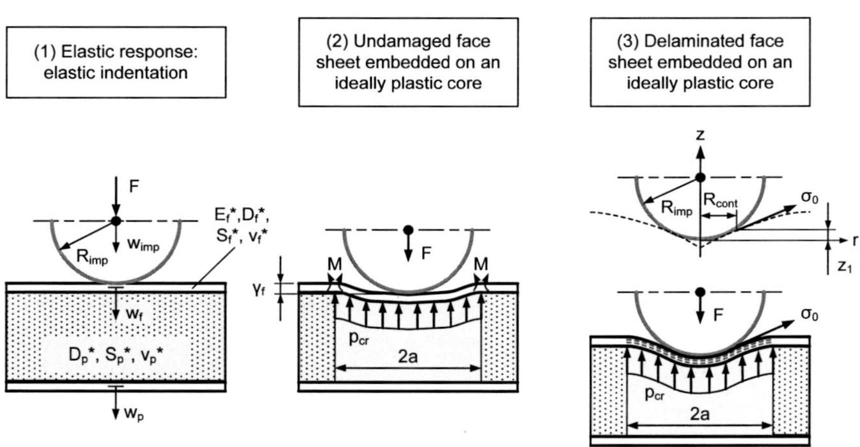

Figure 3.10 – three step indentation model of a sandwich panel developed by Olsson [Ols02]: (1) fully elastic indentation, (2) core crushing leading to ideally plastic core behavior and (3) delamination of the impact face sheet leading to face sheet membrane behavior

In phase (2) the core starts to yield or crush locally below the face sheet thus it is modeled ideally plastic. The face sheet itself remains elastic and does not yet suffer damage.

Olsson [Ols02] assumes for this a finite crushing region with the radius a. In this region the crushed foam core provides a constant reactive pressure p_{cr}, while the outer region of the face sheet is embedded on undamaged foam as elastic foundation. Based on this model two solutions for the local deflection γ_f, the resulting crushing radius a_i and the impactor force F_i are provided. Here the subscript i denotes the solution type. The first solution is based on plate theory (subscript p) while the second solution is based on membrane theory (subscript m) to account for large deflections. The membrane solution can be added to the plate solution as a correction factor but is not considered necessary unless face sheet deflections surpasses about four times the face sheet thickness or delaminations occur. Thus for phase (2) the plate solution is applied for obtaining [Ols02]

$$\gamma_f = \gamma_{cr} + \frac{F_p^2 \bar{a}_p^2}{16\,\pi^2\, p_{cr}\, D_F^* \,(1+\nu_r^*)} \left[(3+\nu_r^*) - \frac{\bar{a}_p^2}{4}(5+\nu_r^*) + 8\frac{\bar{M}}{\bar{F}_p}\right]. \tag{3.25}$$

The dimensionless loads $\bar{M}$ and $\bar{F}_i$ as well as the of the dimensionless crushing zone $\bar{a}_i^2$ in equation (3.25) are normalized by [Ols02]

$$\bar{F}_i = \frac{F_i}{\pi\, p_{cr}}\sqrt{\frac{K_z}{D_f^*}}, \qquad \bar{M} = \frac{M}{p_{cr}}\sqrt{\frac{K_z}{D_f^*}} \qquad \text{and} \qquad \bar{a}_i = \sqrt{\frac{\pi\, p_{cr}\, a_i^2}{F}}. \tag{3.26}$$

The normalized size of the crushing zone $\bar{a}_i^2$ as used in equation (3.25) was tabulated by Olsson [Ols94] and is described as the load fraction that is carried by core crushing [Ols02]. Equation (3.25) can now be solved iteratively using F_{cr} as starting value and stepwise increasing the load until the delamination threshold load F_{dth} is reached. For each load step first $\bar{F}_i$ is determined in order to calculate $\bar{M}$ and $\bar{a}_i^2$. In reference [Ols94] two different values for $\bar{a}_i^2$ are provided depending whether the plate or membrane solution is applied. Depending on the magnitude of the face sheet indentation γ_f, a correction can now be applied to the indentation force of the plate solution F_p to account for the membrane stresses in the face sheet [Ols02]

$$\frac{F}{F_p} = \sqrt{\frac{1+\bar{k}_m \bar{\gamma}_\Delta^2}{[1+(\bar{k}_m - \bar{k}_{mcl})\bar{\gamma}_\Delta^2]^\zeta}} \qquad \text{where} \qquad \bar{\gamma}_\Delta = \frac{(\gamma_F - \gamma_{cr})}{h_f}. \tag{3.27}$$

The parameters $\bar{k}_m$ and $\bar{k}_{mcl}$ are the dimensionless description of the sandwich plate membrane stiffness. These together with the exponent ζ are discussed and tabulated by Olsson in references [Ols94] and [Ols96]. Phase (2) ends as delaminations occur in the impacted face sheet.

In phase (3) local delaminations have initiated and grow in the impacted region of the face sheet in addition to core crushing. Due to the delaminations the face sheet loses its bending stiffness almost completely. Thus the face sheet behavior changes locally from a plate type behavior to a membrane type behavior. For determination of the beginning of phase (3) one has to focus at first on the delaminations. The force F_{dn} necessary for n delaminations in the impacted skin to grow can be determined by [Ols02, Ols06]

$$F_{dn} = \pi\sqrt{\frac{32\, D_f^*\, G_{IIc}}{(n+2)\left(1-\frac{\bar{a}_d^2}{2}\right)}} \qquad \text{where} \qquad \bar{a}_d^2 = \frac{\pi\, p_{cr}\, a_d^2}{F}. \tag{3.28}$$

Here a_d is the radius of the delaminations. The threshold load F_{dth} at which the first delamination starts can now be determined by imposing $n = 1$ and $\bar{a}_d = 0$ [Ols02,Ols06]

$$F_{dth} = \pi\sqrt{\frac{32}{3}\, D_f^*\, G_{IIc}}\,. \tag{3.29}$$

The actual size of the delaminated area can now be determined [Ols02]

$$A_d = \sqrt{\frac{F_{dn}\bar{a}_d^2}{\pi\, p_{cr}}} \qquad \text{where} \qquad \bar{a}_d^2 = 2\left[1 - \left(\frac{F_{dth}}{F_{dn}}\right)^2 \frac{3}{n+2}\right]. \tag{3.30}$$

For determination of n one has to keep in mind that only interfaces between plies with different fiber orientations are subject to delamination. Furthermore delaminations typically have a peanut like shape [Abr97]. Thus an experimentally determined delamination radius a_d is likely to be much larger than the analytically calculated radius as the later assumes a fully delaminated circle. In consequence Olsson uses for monolithic plates a factor of 0.3 to account for the peanut shape in reality versus the full circle in the model [Ols01]. The same can also be applied to delaminations in the face sheet of sandwich structures.

Due to the delaminations in the face sheet, its bending stiffness reduces severely such that membrane behavior will dominate within the delaminated region. Olsson [Ols02] derives a membrane type solution for the sandwich indentation γ_f

$$\gamma_f = \gamma_{cr} + [1 + Cs_1 - (1 + C)s_1^2]w_m + \frac{1}{2}\frac{s_1^2 a_m^2}{R_1}. \tag{3.31}$$

Here w_m is the solution for a concentrated load from [Ols94] while s_1 is the dimensionless contact radius r_1 [Ols02]:

$$w_m = f_w\left(\frac{F_m^2 \bar{a}_m^2}{2\, p_{cr}\, E_{rf}^*\, h_f}\right)^{\frac{1}{3}} \qquad \text{and} \tag{3.32}$$

$$s_1 = \frac{r_1}{a_m} = \frac{-C}{\frac{a_m^2}{w_m R_{imp}} - 2(1+C)}. \tag{3.33}$$

The function f_w is solved in reference [Ols94]. The radii r_1 and R_{imp} are shown in Figure 3.10. In this approach Olsson [Ols02] assumes for the deflection w outside of the contact radius s_1 the shape function

$$w = [1 + Cs_1 - (1 + C)s_1^2]\, w_m\,. \tag{3.34}$$

Thus the three deflection terms in equation (3.31) can be divided into their individual contributions. The deflection γ_{cr} is present at the edge of the crushed region. The shape function w covers the membrane deflection up to the beginning of the contact radius s_1 and the third term adds the contribution of the contact radius s_1 with the impactor.

The indentation load in phase (3) can be determined from its individual contributions similar to the sandwich indentation. Relevant contributions are the force carried by the membrane stresses, the reactive core pressure from equation (3.32) and the contribution from the shear force that drives delamination growth from equation (3.28) [Ols02]

$$F = F_m + F_b(n) = F_m + \pi \sqrt{\frac{32\, D_f^*\, G_{IIc}}{(n+2)\left(1-\frac{\bar{a}_d^2}{2}\right)}}. \tag{3.35}$$

Phase (3) ends with face sheet rupture – tensile fracture of the face sheets – as the membrane stresses exceed the material strength. Olsson [Ols02] derives the membrane stress σ_0 at the edge of the contact radius by

$$\sigma_0 = \frac{p_{cr}\, R_{imp}}{2\, h_f}\left(\frac{1}{s_1^2 \bar{a}_m^2} - 1\right). \tag{3.36}$$

The strain in the contact zone can then be determined by [Ols02]

$$\varepsilon_0 = \sigma_0(1-\nu) = \frac{(1-\nu)\, p_{cr}\, R_{imp}}{2\, h_f}\left(\frac{1}{s_1^2 \bar{a}_m^2} - 1\right). \tag{3.37}$$

Face sheet rupture or rupture occurs at the tip of the impactor as the maximum strain in the face sheet is exceeded. The critical face sheet rupture load F_{rup} can be determined using a simplified approach provided by Olsson and Block [Ols13]

$$F_{rup} = \frac{4\, \pi\, R_{imp}\, h_f\, E_r^*\, \varepsilon_{1t}^2}{(1-\nu_r^*)}. \tag{3.38}$$

Here ε_{1t} is the tensile failure strain of the composite in the main load bearing direction which is typically close to the failure strain of the reinforcing fiber itself. E_r^* and ν_r^* are the elastic properties of an equivalent quasi-isotropic face sheet laminate.

The failure load of equation (3.38) is based on the same approach as equations (3.36) and (3.37) but assumes that within the contact zone the skin carries the entire load of the impactor and neglects the contribution of the crushed core below the contact zone. As the radius of the contact zone r_1 is usually small, this simplification is acceptable. Overall contribution of core crushing to the indentation load is not neglected as the loaded skin redistributes the indentation force via membrane stresses on the crushing zone with radius a. The crushing zone is typically one order of magnitude larger than the contact zone.

3.5 Sandwich failure modes during impact

Impact tests on foam core sandwich structures revealed different failure modes depending on the individual sandwich configuration [Blo11]. The two governing failure modes observed are face sheet rupture initiated by local indentation (Figure 3.11, left) and core shear failure due to a low bending stiffness of the sandwich plate (Figure 3.11, right). A combination of both was observed in cases with high impact energies.

Specimens that fail by core shear may also fail due to face sheet rupture as higher impact energies are applied (Figure 3.11, center). Core shear failure reduces the bending stiffness but does not stop the impacted face sheet from being loaded by membrane stresses which then cause face sheet rupture after core shear failure has already occurred. It is however not possible that a specimen, which initially fails by rupture of the impacted face sheet, later also suffers from core shear failure as the peak impact force is typically reached as face sheet rupture occurs. This out-of-plane load is the driver behind plate bending and subsequently core shear stresses that cause core shear failure.

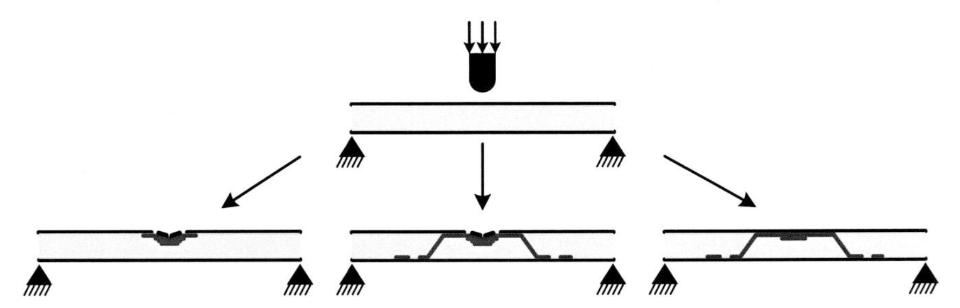

Figure 3.11 – sandwich impact failure modes: Face sheet rupture (left), co-occurrence of face sheet rupture and core shear failure (center) and core shear failure (right)

To determine which failure mode is going to occur, the loads that cause either failure mode have to be compared. Equation (3.38) describes the face sheet rupture load which is the peak out-of-plane force of an impactor of a given diameter R_{imp} on a sandwich plate. Olsson and Block [Ols13] determined the maximum shear stress in the core to

$$\bar{\tau}_c = \frac{\varepsilon_{1t}}{h_c}\sqrt{\frac{R_{imp}\, h_f\, E_r^*\, \bar{a}^2\, p_{cr}}{(1-\nu_r^*)}} \,. \tag{3.39}$$

Combining equations (3.38) and (3.39), the impactor load at core shear failure F_{cs} can be determined. Here $\bar{\tau}_c$ is replaced by the core shear strength $\hat{\tau}_c$

$$F_{cs} = \frac{4\,\pi\,\hat{\tau}_c^2\, h_c^2}{\bar{a}^2\, p_{cr}} \qquad \text{where} \qquad \bar{a}^2 \leq 0.8 \,. \tag{3.40}$$

The dimensionless radius $\bar{a}^2$ of the core crush region is here interpreted as a load reduction factor that describes the contribution of global membrane stresses to the load on the impactor [Ols13].

Transition between the two failure modes can now be established by equating the critical forces F_{rup} and F_{cs} as described by equations (3.38) and (3.40)

$$\frac{R_{imp}\, h_f\, E_r^*\, \varepsilon_{1t}^2}{(1-\nu_r^*)} = \frac{\hat{\tau}_c^2\, h_c^2}{\bar{a}^2\, p_{cr}} \,. \tag{3.41}$$

From the perspective of a designer it is important to know the minimum foam core thickness $h_{c,min}$ at a given face sheet thickness h_f in order to avoid core shear failure. This can be achieved by reorganizing equation (3.41) to

$$h_{c,min} = \frac{\varepsilon_{1T}}{\hat{\tau}_c} \sqrt{\frac{R_{imp}\, h_f\, E_r^*\, \bar{a}^2\, p_{cr}}{1-\nu_r^*}} \qquad \text{where} \qquad \bar{a}^2 \le 0.8\ . \tag{3.42}$$

This was now solved for the material combination of CFRP and Rohacell 71RIST. Tensile strain at face sheet rupture ε_{1T} is assumed to be equal to the maximum tensile strain of the HTS carbon fiber. The impactor radius R_{imp} was chosen to 12.7 mm. The results are shown in Figure 3.12. As described by equation (3.42) the relationship between the two parameters h_f and h_c is not linear. The material properties used for the failure mode map are listed in Table 3.1.

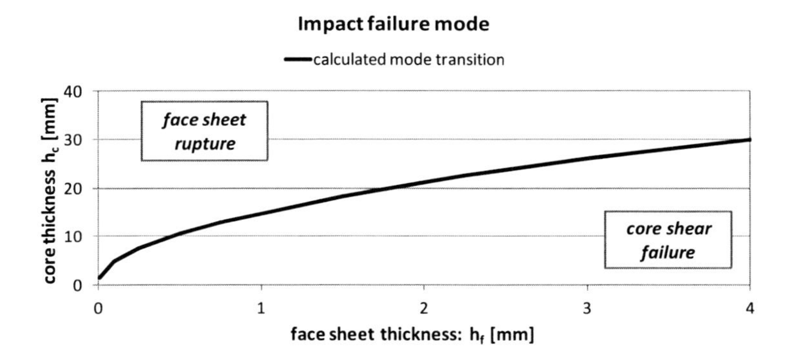

Figure 3.12 – failure mode map for low-velocity impact of foam core composite sandwich plates

Table 3.1 – sandwich material properties used for creation of the failure mode map

E_r^* [MPa]	ν_r^* [-]	ε_{1T} [%]	p_{cr} [MPa]	$\hat{\tau}_c$ [MPa]	R_{imp} [mm]	$\bar{a}^2$ [-]
48850	0.307	1.8	1.9	1.4	12.7	0.8

3.6 Chapter summary

This chapter focuses on describing the impact response of composite foam core sandwich structures based on experimental knowledge from the literature and analytical methods. In engineering applications the impact response is often reduced to an impact resistance design requirement and thus concentrates on the damage created during foreign object collisions. It is typically treated as part of the design requirement damage tolerance.

The impact response of composite sandwich plates can be divided into the structural response of the sandwich plate and the local sandwich indentation response. The structural response can be classified into three major types: low velocity impact, high velocity impact and ballistic impact. They differ from one another on how stress wave propagation in the horizontal and vertical plate direction influences the structural behavior. Low velocity impact behavior – also referred to as quasi-static or large mass impact – is characterized by the absence of any influence of wave propagation on the structural behavior and is thus controlled by the boundary conditions of the impacted plate.

The structural behavior of a plate during high velocity impact – also referred to as small mass impact – is controlled by stress wave propagation in the plate horizontal direction while boundary conditions and propagation in the vertical direction have no or little influence. In contrast ballistic impact behavior is predominantly influenced by stress wave propagation in the vertical plate direction.

This work concentrates on low velocity impact. Distinction between low and high velocity impact can be performed using the mass ratio of the impactor and the impacted plate. For low velocity impact the mass of the impactor must be at least two times bigger than the mass of the impacted plate. A decision tree for the characterization of the structural response by Chai and Zhu [Cha11] is provided in Figure 3.7. Models for low and high velocity structural response are discussed.

The contact behavior of composite sandwich structures is discussed together with a three step sandwich indentation model from Olsson [Ols02]. This model provides a good understanding of the mechanical processes that occur during the impact. In the first step the behavior is fully elastic until core crushing occurs. In the second step the core crushes and is modeled with ideally plastic behavior. The second step ends as additionally to core crushing also delaminations occur in the face sheet. During the third step of indentation the face sheet loses due to the delaminations its entire bending stiffness and is completely loaded by membrane stresses and backed by the continuously crushing core. The third step ends as either the sandwich face sheet ruptures or the foam core fails due to shear loads. Failure criteria from Olsson and Block [Ols13] are presented. Based on these criteria the transition between either of the two failure modes is established and demonstrated as a failure mode map shown in Figure 3.12.

The failure mode map is intended as a guideline for designers. It is based on the design values face sheet thickness and core thickness. The failure mode map shows the transition line between face sheet rupture and core shear failure for a selected material combination. From this map the designer can select the thicknesses that will lead to either of the discussed failure modes.

4 Experimental investigation of the impact response

4.1 Test matrix and setup

Experimental investigations were separated into two series of parametric impact tests. The test series are labeled RT and Frost which corresponds with the environmental test conditions. Test series RT was conducted at 20 °C equivalent to room temperature (RT) while the impacts of test series Frost were performed at -55 °C.

The impact tests were conducted for understanding the underlying mechanisms that cause damage in sandwich structures during impact loading. Additionally information on damage visibility and size was collected. For all specimens impact testing parameters were kept constant except for the applied impact energy and the environmental conditions as described previously. Impact energy varied between 12 J and 90 J depending on the thickness and strength of the sandwich configuration. This aims at finding the critical energy threshold required for initiation of face sheet rupture or core shear failure. Sandwich configurations varied only in their face sheet and core thicknesses.

The investigated sandwich panels are manufactured using dry triaxial non-crimp fabrics (NCF) made of Toho Tenax HTS carbon fibers. The triaxial NCFs are produced by the company Saertex GmbH & Co. KG of Saerbeck, Germany with a stacking sequence of [45°/0°/135°]. The area weight is 135 g/m² of carbon fiber per layer and 4 g/m² PES sewing thread across all layers, in total 409 g/m². This results in a nominal ply thickness of 0.125 mm for 60% fiber volume content (FVC). The complete NCF thus has a nominal thickness of 0.375 mm. HexFlow RTM 6 resin was used as matrix for the face sheets while the foam core was made of Rohacell 71RIST PMI foam with a nominal density of 75 kg/m³. Material properties are summarized in appendix B.

Four different symmetrical stacking sequences $[((45°/0°/135°)_s)_n]$ with n varying from 1 to 4 were used for the face sheets. Combined with five different core thicknesses that vary from 6.5 mm to 35.5 mm this results in a total of 20 possible sandwich configurations. Of these only 14 were built, which was sufficient for describing the relevant failure modes face sheet rupture and core shear failure. Three specimens were built of each configuration leading to a total 42 available specimens for series RT. Additionally 4 configurations and thus 12 specimens were fabricated for the Frost series. Test matrices and the applied naming convention for the sandwich configurations are summarized in Figure 4.1 and Figure 4.2.

Sandwich panels of size 1150 x 500 mm were fabricated and subsequently cut into 350 x 400 mm specimens. The specimens were manufactured using the modified vacuum infusion (MVI) technology. The full sandwich stacking sequence was built on a flat steel tooling plate placing the dry NCFs and the foam core as shown previously in Figure 2.4 in section 2.2. For infusion and curing the complete setup was placed into a hot air oven. Infusion of the resin was performed at 120 °C while curing took place at 180 °C.

RT tests		h_{core} = 6.5 mm	h_{core} = 10.0 mm	h_{core} = 16.3 mm	h_{core} = 25.7 mm	h_{core} = 35.5 mm	
2 NCF plies	h_{face} = 0.75 mm	H1a	H1b	H1c	-	-	(3 heights)
4 NCF plies	h_{face} = 1.5 mm	6.5	10.0	16.3	25.7	35.5	(5 heights)
6 NCF plies	h_{face} = 2.25 mm	-	-	H2a	H2b	H2c	(3 heights)
8 NCF plies	h_{face} = 3.0 mm	-	-	H3a	H3a	H3b	(3 heights)
Specimen count		6	6	12	9	9	42

Figure 4.1 – test matrix of RT impact tests (20 °C) and naming of sandwich configurations

Frost tests		h_{core} = 6.5 mm	h_{core} = 10.0 mm	h_{core} = 16.3 mm	h_{core} = 25.7 mm	h_{core} = 35.5 mm	
2 NCF plies	h_{face} = 0.75 mm	-	-	-	-	-	(0 heights)
4 NCF plies	h_{face} = 1.5 mm	-	-	-	25.7-1.5	35.5-1.5	(2 heights)
6 NCF plies	h_{face} = 2.25 mm	-	-	-	25.7-2.25	35.5-2.25	(2 heights)
8 NCF plies	h_{face} = 3.0 mm	-	-	-	-	-	(0 heights)
Specimen count		0	0	0	6	6	12

Figure 4.2 – test matrix of Frost impact tests (-55 °C) and naming of sandwich configurations

Impact testing was conducted on an Instron CEAST 9350 drop weight impact testing machine at the Fraunhofer Institute for Mechanics of Materials in Halle, Germany. The test machine is equipped with a load cell for measuring force history and related acceleration and displacement data during the impact. It is also equipped with a mechanism to prevent a second impact due to rebound of the impactor. This mechanism catches the impactor prior to hitting the test specimen a second time.

During the impact the following data was recorded:

- contact force history,
- impactor displacement, velocity and acceleration history,
- dissipated energy and
- velocity of the impactor when it hits the panel.

Impact testing was performed with a 25.4 mm diameter hemispherical steel impactor. The impact energies range from 12 J to 90 J. The weight of the impactor was kept constant at 3.075 kg for impact energies up to 50 J. For tests above 50 J the weight was increased up to 5.075 kg in order to limit the impact velocity below 6 m/s.

The test specimens were secured in a picture frame fixture, which clamps the panel around its edges and provides an unsupported open window in the center. The fixture has a size of 450 x 500 mm. The open window in the center is of size 250 x 300 mm providing 50 mm support length along the edges. The specimen itself has a size 350 x 400 mm as shown in Figure 4.3. The laminate 0° orientation is the specimens long edge. The test fixture and resulting boundary conditions are shown in Figure 4.4. The specimen is enclosed within the picture frame fixture using bar clamps that are seized against a metal support plate.

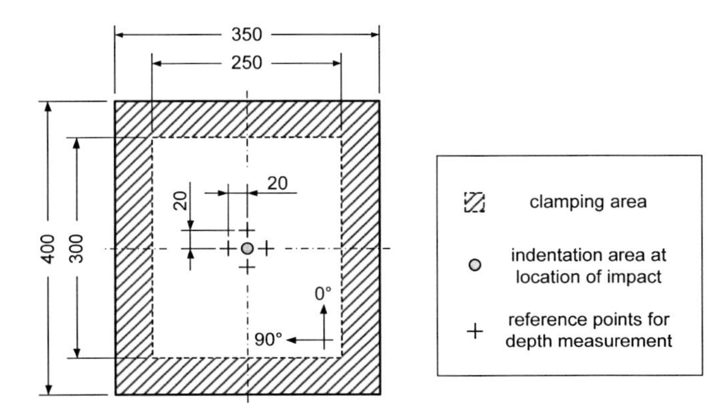

Figure 4.3 – sandwich impact test specimen

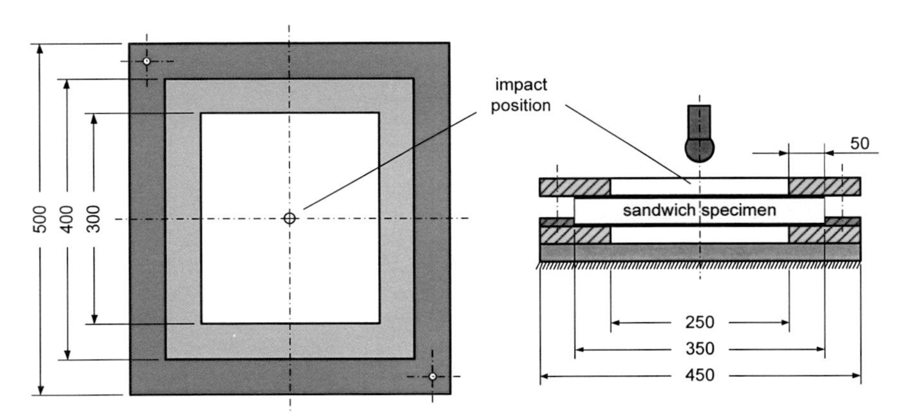

Figure 4.4 – picture frame test fixture (bar clamps not shown, dimensions in mm)

The impacted sandwich panels were analyzed non-destructively using air coupled through-transmission ultrasound and residual dent measurements. Residual dent measurements were performed directly after impact testing in accordance to the procedure described in AITM 1-0010 [Air05]. The effect of relaxation on the dent depth due to e.g. moisture absorption was thus not taken into account. In total five depth measurements per specimen were made using a caliper. One measurement was taken at the point of impact while the other four measurements were taken at reference points 20 mm from the point of impact each in a different direction. Figure 4.3 shows this arrangement. The effective dent depth was then determined by subtracting the average value at the reference points from the depth measurement at the point of impact.

For non-destructive inspection (NDI) of the sandwich specimens C-scans based on air coupled ultrasound had been selected. This was performed at the German Aerospace Center (DLR) in Braunschweig, Germany. Air coupled ultrasound requires lower ultrasound frequencies than water coupled ultrasound due to higher damping and reflection losses of the ultrasound. Typically monolithic CFRP is tested non-destructively with ultrasound in the range of 2 to 5 MHz while air coupled ultrasound uses frequencies in the range from 50 to 300 kHz. Consequently the planar resolution of air-coupled ultrasound is, due to the larger wave length of the sound waves, lower than conventional water coupled ultrasound.

Depending on the specimen a frequency of 75 kHz or 120 kHz was used. As attenuation increases with specimen thickness, specimens with a core thickness $h_{core} > 20$ mm were tested with a frequency of 75 kHz while thinner specimens could be tested with 120 kHz. Air coupled ultrasound is capable of detecting both face sheet damage within the CFRP as well as foam damage such as debondings or cracks. It is however not possible to determine the specific damage type with absolute certainty as the ultrasound through transmission mode is not capable of recording the time of flight of the sound signal and can thus not provide information on the depth of detected damage within the specimen. Thus only total damage size could be determined as shown exemplary for two specimens in Figure 4.5. In case of the left example the damage is limited locally to the face sheet and core while the right specimen depicts a much larger damage. Sectioning of selected specimens was then performed to investigate the damage types and confirm their size.

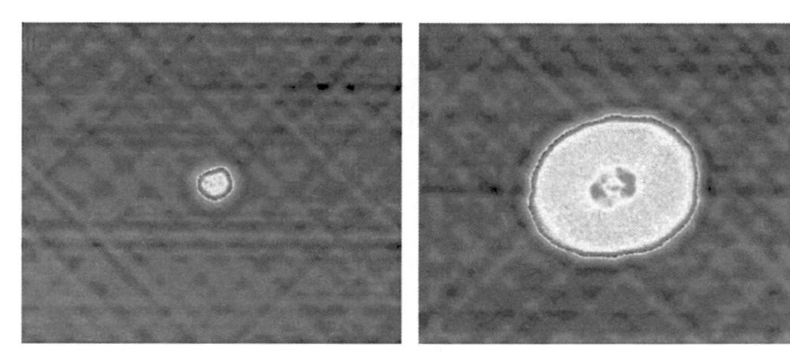

Figure 4.5 – NDI results (air coupled ultrasound, C-scan):
20°J impact on specimens RT_H1c_P1 (left) and RT_H1b_P1 (right) [Fri11]

4.2 Test results at room temperature

The test results can be divided into impact damage and information about the impact process such as e.g. the force history. First typical impact damages will be described and analyzed using NDI images and sectioned test specimens. After this information about the impact process itself will be presented and investigated.

Generally two different appearances of impact damage were observed that could be related to the sandwich failure modes that were previously discussed in section 3.5. The first appearance includes face sheet cracks and core crushing damage below the point of impact and can be related to the indentation problem and face sheet rupture. The second sandwich failure mode observed is dominated by core shear cracks and rear face sheet delaminations, which can be connected to the core shear failure mode. Here damage occurs due to shear loads in the core which leads to a circular shear crack that extends from the core crushing zone at an angle of 45° to 60° relative to the face sheet through the core. The crack then continues along the rear face sheet interface leading to a hat shaped appearance. Figure 4.6 shows sectioned test specimens subject to face sheet rupture while Figure 4.7 shows specimens with core shear failure. Scratch marks in the foam core are a product of the sectioning process and thus of no significance. The shear failure does not appear as a perfect crack but is instead a mixture of a band of deformed cell walls and stepwise cracks. A schematic description of both failure modes is given in Figure 3.11.

Simultaneous occurrence of both failure modes was also observed but limited to configurations that experience core shear failure already at lower impact energies but are instead hit with a relatively large energy. Here the impact energy was large enough to first initiate core shear failure but later also face sheet rupture. Finally all specimens were investigated with air coupled ultrasound as NDI method. Figure 4.8 shows C-scans of all four ply specimens together with the applied impact energy and measured dent depth. The test results of the specimens with two, six and eight plies of NCF are presented in appendix C.1.

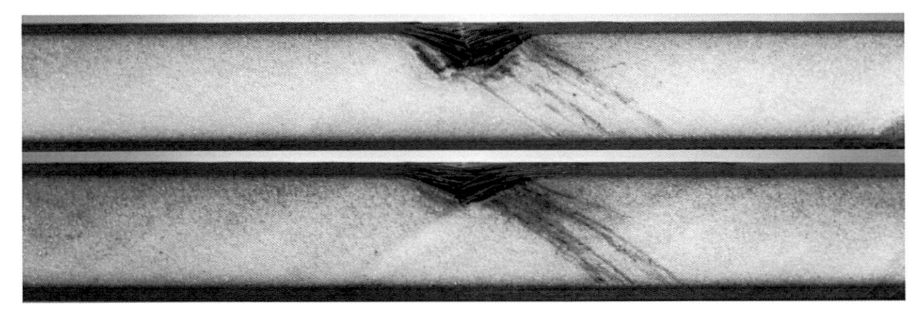

Figure 4.6 – failure mode face sheet rupture observed in specimens RT_H2b_P3 (top) and RT_H3b_P3 (bottom) subject to 90 J impacts

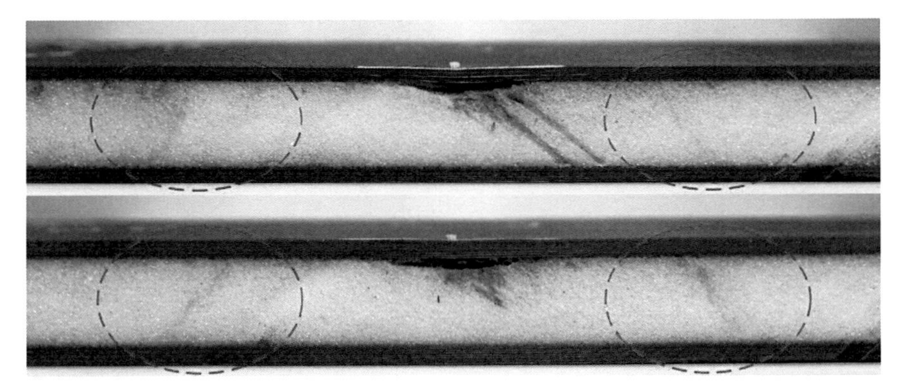

Figure 4.7 – failure mode core shear failure observed in specimens RT_H2a_P2 (top) and RT_H3a_P2 (bottom) subject to 50 J impacts

h_{face}=1.5 mm				
h_{core}=6.5 mm	h_{core}=10.0 mm	h_{core}=16.3 mm	h_{core}=25.7 mm	h_{core}=35.5 mm
RT_6.5_P3 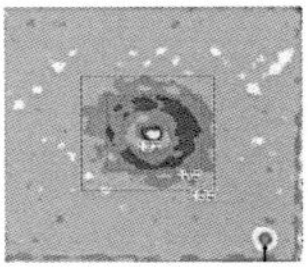 h_{dent}=0.25 mm E_{imp}=12 J	RT_10.0_P3 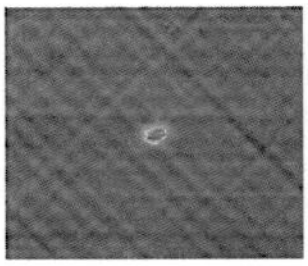 h_{dent}=0.15 mm E_{imp}=12 J	RT_16.3_P1 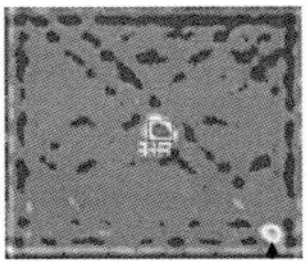 h_{dent}=0.13 mm E_{imp}=20 J	RT_25.7_P1  h_{dent}=1.0 mm E_{imp}=20 J	RT_35.5_P1 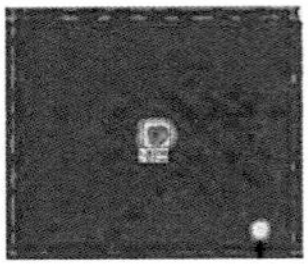 h_{dent}=2.86 mm E_{imp}=35 J
RT_6.5_P1 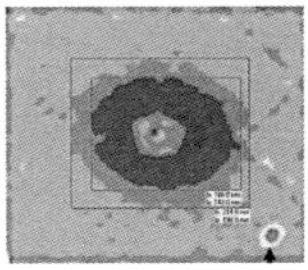 h_{dent}=0.1 mm E_{imp}=20 J	RT_10.0_P1 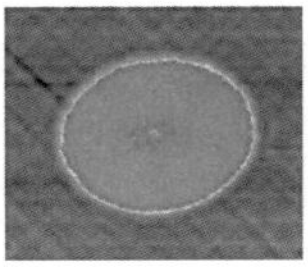 h_{dent}=0.25 mm E_{imp}=20 J	RT_16.3_P2 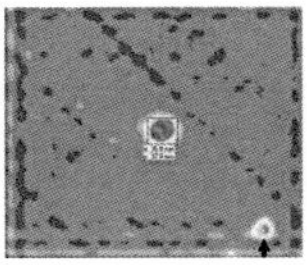 h_{dent}=1.85 mm E_{imp}=35 J	RT_25.7_P2 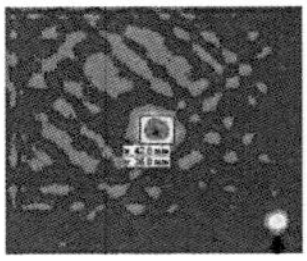 h_{dent}=3.11 mm E_{imp}=35 J	RT_35.5_P2  h_{dent}=6.68 mm E_{imp}=50 J
RT_6.5_P2 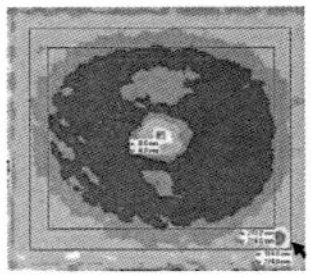 h_{dent}=0.91 mm E_{imp}=35 J	RT_10.0_P2 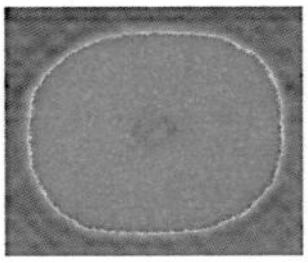 h_{dent}=0.95 mm E_{imp}=35 J	RT_16.3_P3 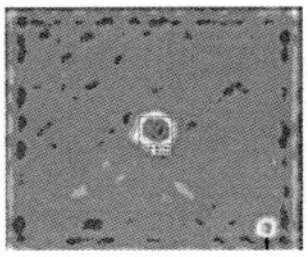 h_{dent}=3.23 mm E_{imp}=50 J	RT_25.7_P3 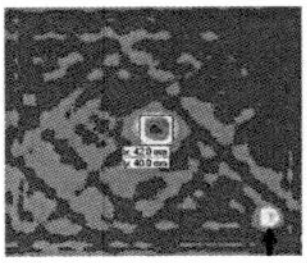 h_{dent}=6.54 mm E_{imp}=50 J	RT_35.5_P3  h_{dent}=24.2 mm E_{imp}=90 J

Figure 4.8 – NDI result of RT impact specimens with 4-ply face sheets (1.5 mm thickness) [Bie11][Fri11]

In order to quantify the severity of the damages, their size was determined based on the air coupled TTU C-scans. Damage size is dominated by core damage as it typically extends across a greater area than the face sheet damage due to core crushing or shear cracks. In Figure 4.9 all damage sizes of the RT tests are plotted against the impact energy. Here damage size is given as the planar damage diameter using the average value in the sandwich x- and y-directions.

Figure 4.9 shows two trends. The first groups damage sizes up to 50 mm in diameter. Here the damage size is more or less independent of the impact energy. The second trend covers larger damage sizes of 100 mm in diameter or more. These start at about 100 mm for impact energies of 12 J and increase up to about 300 mm for impact energies of 50 J. The two trends depict the result of the two discussed failure modes inside the sandwich.

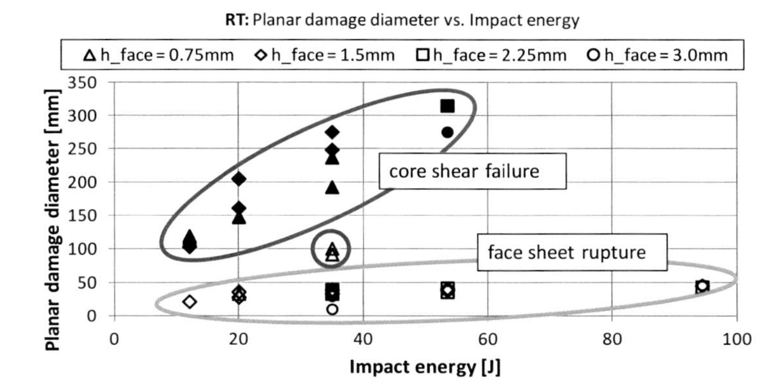

Figure 4.9 – RT impact tests: Damage size vs. impact energy;
empty symbols: Face sheet rupture, filled symbols: Core shear failure

The first observed trend is face sheet rupture (green circle) as the local failure mode that is initiated by face sheet membrane stresses and compressive stresses on the core. Consequently damage is confined to the surrounding of the impactor and its diameter is limited to about twice the diameter of the impactor. The second trend is the core shear failure mode (red circle), which leads to significantly larger damages. Here damage size increases noticeably as more impact energy is applied.

There is only one result in Figure 4.9 which does not fit into either of the two trends. This is the sandwich configuration H1c with a 16.3 mm core and 0.75 mm face sheets at 35 J impact energy (blue circle). At 20 J impact energy the specimen experiences face sheet rupture and core crushing and thus fits well into the first trend. The more powerful impact energy of 35 J causes the impactor to penetrate deeply into the foam core crushing large amounts of core material and pushing this material sideways. As the impactor approaches the rear face sheet this causes local shear cracks and debonding of the rear face sheet as shown in Figure 4.10. This leads to a greater damage size of about 100 mm in diameter.

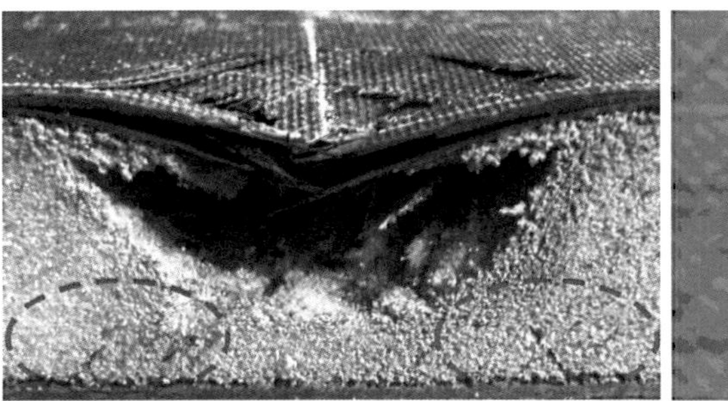
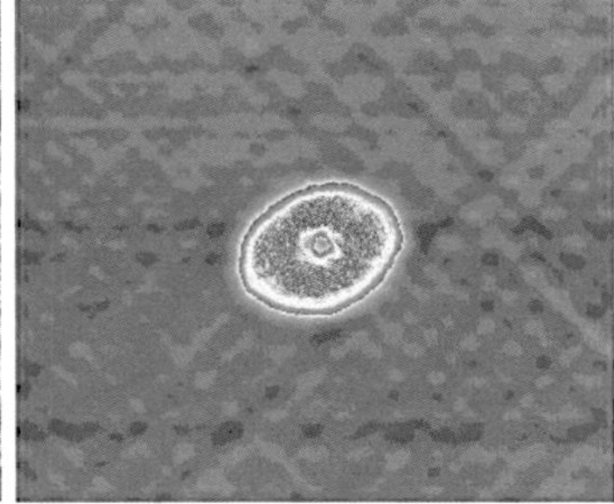

Figure 4.10 – face sheet rupture with crushing induced local shear cracks in specimen RT_H1c_P3 due to a 35 J impact at room temperature: Sectioned specimen (left) and NDI image [Fri11] (right)

Visibility of impact damage is – besides secondary factors such as lighting, experience of the inspector, etc. – governed by residual indentation or simply dent depth [Baa09]. Figure 4.11 plots dent depth against impact energy on a logarithmic scale. This shows that face sheet thickness and impact energy have a noticeable effect on the residual dent depth. It can be summarized that thicker face sheets are subject to lower dent depths and thus damage visibility when compared to thinner face sheets. Higher impact energy creates greater dent depths while the sandwich failure mode has less influence. Core shear failure leads to slightly lower dent depth as energy is consumed in the core which can be noticed in Figure 4.11 when comparing specimens with the same face sheet thickness.

Ideally damage size and visibility can be correlated with one another. Figure 4.12 thus plots residual indentation depth against planar damage size. Similar to Figure 4.9 the results group in agreement with the sandwich failure mode. When comparing only face sheet rupture (green circle), damage size and visibility can be correlated. In case of an indentation depth no larger than 1 mm, the greatest damage size measured is about 40 mm. Increasing indentation depth and thus damage visibility leads to slightly larger damage sizes of up to 50 mm. In contrast core shear failure (red circle) leads to damage sizes of up to 320 mm at the same level of visibility described by an indentation depth of 1 mm.

Figure 4.13 presents exemplary damage visibility of the specimens with 2-ply face sheets (h_{face} = 0.75 mm) only. Here a clear trend is visible that thicker cores promote a greater dent depth and thus better damage visibility. At 35 J impact energy for instance, the specimen with 6.5 mm core thickness has a dent depth of 3.5 mm, while the specimen with 10.0 mm core thickness already has a dent depth of 6.5 mm. This increases to 12.5 mm for the specimens with a 16.3 mm core. In this case the thinner specimens experienced core shear failure while the thicker specimen did not. This trend can also be observed for specimens with 4, 6 and 8-ply face sheets as shown in appendix C.1.

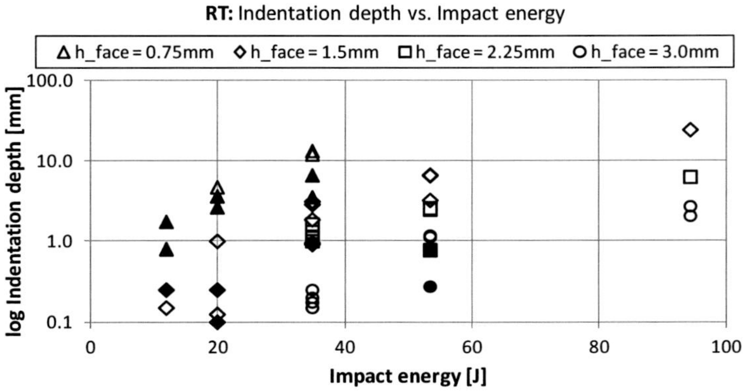

Figure 4.11 – RT impact tests: Indentation depth vs. impact energy;
empty symbols: Face sheet rupture, filled symbols: Core shear failure

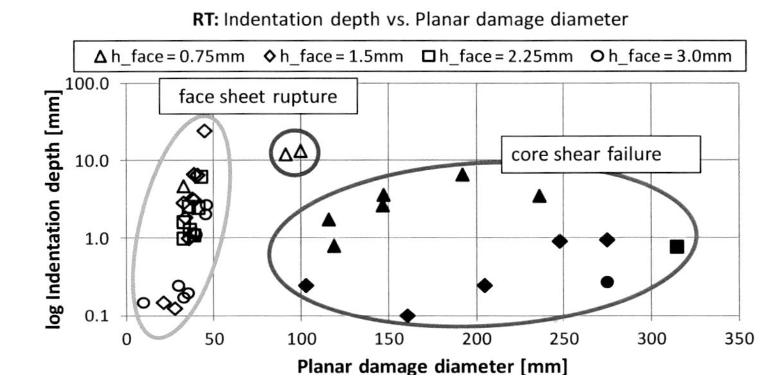

Figure 4.12 – RT impact tests: Indentation depth vs. planar damage diameter; empty symbols: Face sheet rupture, filled symbols: Core shear failure

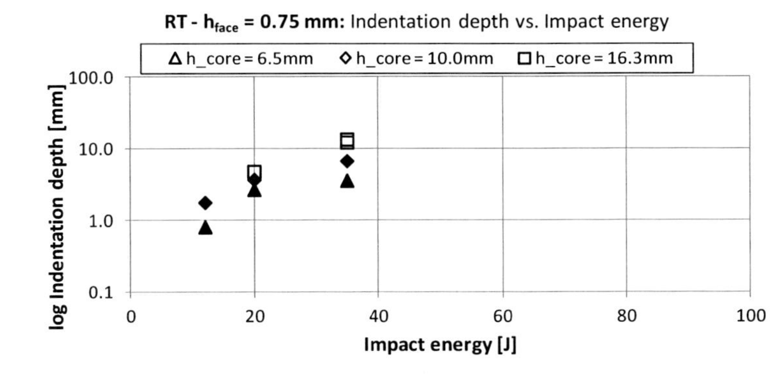

Figure 4.13 – RT impact tests: Indentation depth vs. impact energy of specimens with 2-ply face sheets only; empty symbols: Face sheet rupture, filled symbols: Core shear failure

For a more refined investigation of the effect of the failure mode, attention is put on the specimens with 6-ply face sheets (h_{face} = 2.25 mm) and 8-ply face sheets (h_{face} = 3.0 mm). Detailed impact test results of these specimens are provided in appendix C.1 but are also included in Figure 4.11 and Figure 4.12. For both face sheet thicknesses the specimens with a 16.3 mm thick core only experience core shear failure due to a 50 J impact while there is only minor damage observed after a 35 J impact. In case of the 6-ply face sheet specimen the dent depth actually declines slightly from about 1 mm for the 35 J impact to 0.8 mm for the 50 J impact while in case of the 8-ply face sheet specimen dent depth increases only marginally from about 0.2 mm for the 35 J impact to 0.28 mm for the 50 J impact. This is in contrast to the other specimens with a thicker core. Here the observed dent depth increases sharply by a factor of two to five as the impact energy is raised from 35 J to 50 J.

Summarizing this it can be stated that the core shear failure mode has a detrimental effect on damage visibility. As the discussed results are based on only one or two specimens per impact energy, the effect of statistic scatter shall however be kept in mind. Thus no statement regarding specific threshold values for BVID and VID is made.

The impact response of the sandwich specimens is characterized by the force history of the impactor and the corresponding displacement data. These can be evaluated to find specific points of interest during the impact event. Figure 4.14 shows the results of four 4-ply specimens with different core thicknesses subject to a 20 J impact. As one may recall, the specimens with a core thickness of 6.5 mm and 10.0 mm were subject to core shear failure, while the specimens with 16.3 mm and 25.7 mm were not. The specific results of the impact tests are also summarized in Table 4.1.

The different load curves behave similar until 1000 to 1500 N. Here all curves experience the first noticeable load drop leading to the conclusion that minor damage initiates already at this point. From here on the curves differentiate depending on their core thickness which largely determines the sandwich plate bending stiffness.

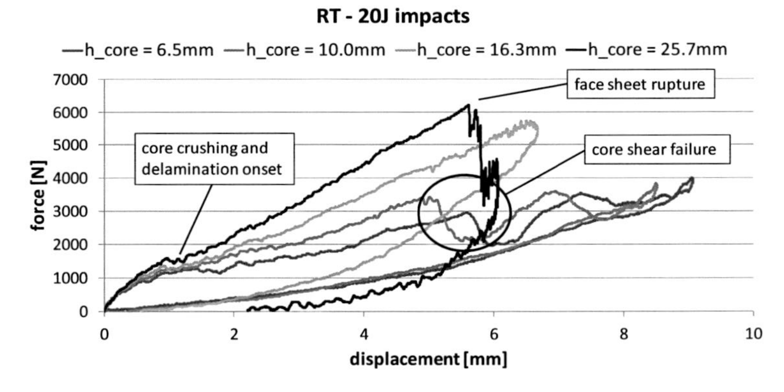

Figure 4.14 – impact response of different 4-ply specimens subject to 20 J impacts; specimen core thickness varies from 6.5 mm to 25.7 mm

Table 4.1 – experimental results of 20 J impact tests on 4-ply (h_{face} = 1.5 mm) specimens

configuration / specimen	core thickness [mm]	peak force [N]	max. displacement [mm]	dent depth [mm]	absorbed energy [%]	damage diameter [mm]
RT_6.5_P1	6.5	3758	9.08	0.1	38	161
RT_10.0_P1	10	4030	8.52	0.25	46	205
RT_16.3_P1	16.3	5731	6.67	0.13	49	28
RT_25.7_P1	25.7	6218	6.07	1.0	85	36

A closer look to the force vs. displacement plot reveals also a more minor stiffness drop between 500 and 750 N. Following the analytical indentation model the impactor force at onset of core crushing is described by equation (3.23). Using the thick core assumption F_{cr} becomes 413 N. As F_{cr} is sensitive to the rate dependent core crushing stress, a greater experimental value of 500 to 750 N is considered as realistic. Similarly equation (3.29) describes the impactor force required for superseding the delamination threshold force F_{dth} and becomes 1498 N. Here the analytical equation slightly overestimates the experimentally observed load drop at 1000 to 1500 N. Table 4.2 summarizes experimental results and analytical results. The analytical results were determined using the material properties from appendix B.

Table 4.2 – analytical and experimental results of 20 J impacts on 4-ply (h_{face} = 1.5 mm) specimens

configuration / specimen	failure mode	F_{cr} [N]	F_{dth} [N]	F_{cs} [N]	F_{rup} [N]	failure force [N]	peak force [N]
RT_6.5_P1	core shear failure	413	1498	685	5462	2985	3758
RT_10.0_P1	core shear failure	413	1498	1620	5462	3402	4030
RT_16.3_P1	none	413	1498	4305	5462	-	5731
RT_25.7_P1	face sheet rupture	413	1498	10703	5462	6218	6218

The two thinner specimens RT_6.5_P1 and RT_10.0_P1 (h_{core} = 6.5 mm / 10.0 mm) shown in Figure 4.14 are subject to core shear failure and experience a major load drop between 3000 and 3500 N. This drop has a magnitude of about 1000 N and sustains for about one to two millimeter until the previous load level is restored. The core shear failure force F_{cs} from equation (3.40) becomes 685 N for specimen RT_6.5_P1 and 1620 N for RT_10.0_P1. This is noticeably lower than 3000 N and 3500 N observed during the impact tests. There is no complete explanation to this but the RIST foam has some ductility before it fails. Thus a local overload may not immediately lead to core shear cracks. Figure 4.7 revealed in sectioned test specimens that core shear failure takes place partially by local plasticity of the cell walls and stepwise cracking across multiple foam cells.

After the occurrence of core shear failure the load curves of the specimens RT_6.5_P1 and RT_10.0_P1 rise slowly to a level of about 4000 N where peak displacement is reached. Unloading of the sandwich plate takes place very gradually. The dent depths of 0.1 mm and 0.25 mm indicate that face sheet rupture is not present. The low amount of 38% and respectively 46% absorbed energy reveal that a large part of the impact energy was stored elastically and released on impactor rebound which agrees well with a small hysteresis loop.

The thickest specimen RT_25.7_P1 (h_{core} = 25.7 mm) shows a different behavior. Here the force rises from about 1500 N to 6000 N almost linearly but then drops off sharply at about 6200 N while experiencing significant scatter which indicates face sheet rupture. This corresponds with a greater residual dent depth of 1 mm and the large amount of 85% absorbed energy. In contrast the impact response of specimen RT_16.3_P1 (h_{core} = 16.3 mm) shows no signs of core shear failure or face sheet rupture. The impactor reaches a peak force of about 5700 N with some scatter occurring at this level. This can be understood as an indication that limited fiber cracking occurs in the face sheet and thus face sheet rupture is imminent but did not yet occur. The low dent depth of only 0.13 mm further indicates that no face sheet rupture occurred.

According to equation (3.38) the impactor force required for face sheet rupture F_{rup} is in case of the four discussed configurations with 1.5 mm thick face sheets equal to 5462 N. By contrast the core shear failure force F_{cs} from equation (3.40) is for configuration RT_16.3 only 4305 N and increases for configuration RT_25.7 to 10703 N. In case of the thicker configuration RT_25.7 F_{cs} is significantly larger than the face sheet rupture force of 5462 N which agrees with the observed failure mode during the impact test. In case of configuration RT_16.3 this is however not the case. Here F_{cs} is smaller than F_{rup}. Thus – following the analytical failure criterion – configuration RT_16.3 should be subject to core shear failure provided that the impact energy is large enough. In the experiment however not even a 50 J impact triggered core shear failure as shown in Figure 4.8.

Figure 4.15 shows additionally the impact response of specimens RT_H2a_P2, RT_H2b_P2 and RT_H2c_P2 which were subject to 50 J impacts. The specimens are made of 6 NCF plies with a nominal face sheet thickness of 2.25 mm and have core thickness of 16.3 mm, 25.7 mm and 35.5 mm. Results of the analytical failure criteria are summarized together with experimental results in Table 4.3 and Table 4.4.

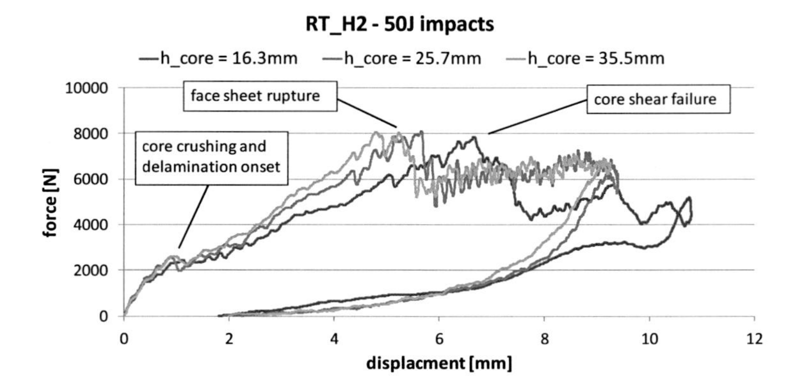

Figure 4.15 – impact response of different 6-ply specimens subject to 50 J impacts; specimen core thickness varies from 16.3 mm to 35.5 mm

Table 4.3 – experimental results of 50 J impact tests on 6-ply (h_{face} = 2.25 mm) specimens

configuration / specimen	core thickness [mm]	peak force [N]	max. displacement [mm]	dent depth [mm]	absorbed energy [%]	damage diameter [mm]
RT_H2a_P2	16.3	3758	10.79	0.78	65	315
RT_H2b_P2	25.7	4030	9.40	2.48	70	42
RT_H2c_P2	35.5	5731	9.26	2.58	69	36

Table 4.4 – analytical and experimental results of 50 J impacts on 6-ply (h_{face} = 2.25 mm) specimens

configuration / specimen	failure mode	F_{cr} [N]	F_{dth} [N]	F_{cs} [N]	F_{rup} [N]	failure force [N]	peak force [N]
RT_H2a_P2	core shear failure	929	2752	4305	8192	7859	7859
RT_H2b_P2	face sheet rupture	929	2752	10703	8192	8098	8098
RT_H2c_P2	face sheet rupture	929	2752	20421	8192	8060	8060

Focusing at first on core crushing and delamination onset a thicker face sheet of 2.25 mm leads to significantly higher loads for the onset of damage. Following the analytical approach of equations (3.23) and (3.29), F_{cr} and F_{dth} occur at 929 N and 2752 N. The force vs. displacement plot shows a first minor stiffness drop at around 800 N, a more noticeable second stiffness drop at about 1500 N and a first load drop at about 2500 N. The origin of the stiffness drops may not be traced accurately as they may interfere with oscillations due to dynamic effects. It is however noted that the second and more noticeable stiffness drop at about 1500 N coincides with the analytically determined core crushing load. The first load drop at about 2500 N agrees reasonably with the analytically determined force for delamination onset. Here the experimental force is smaller than the calculated force. These trends agree with those trends previously observed with the 1.5 mm face sheet specimens. For confirmation of the analytical model however static indentation tests up to predefined force levels and subsequent destructive sectioning or none destructive testing will be required.

After this first observable load drop the impactor force then increases for all three specimens up to about 8000 N where it drops off sharply. Impactor displacement of the peak load decreases with increasing core thickness and thus in agreement with specimen bending stiffness. The force of the thinner RT_H2a_P2 specimen however drops off to loads around 4000 to 5000 N while the thicker RT_H2b_P2 and RT_H2c_P2 specimens continue loading at 6000 to 7000 N. NDI showed that specimen RT_H2a_P2 experienced core shear failure while the thicker specimens failed by face sheet rupture as shown in appendix C.1. The analytical failure forces for face sheet rupture F_{rup} is according to equation (3.38) for all specimens equal to 8192 N and thus agrees well with the experimental results of the thicker specimens RT_H2b_P2 and RT_H2c_P2. The thinner specimen RT_H2a_P2 also shows local oscillations prior to the peak load which indicates local fiber cracking as onset of face sheet rupture. The specimen however failed by core shear failure instead. The analytical core shear failure load F_{cs} is according to equation (3.40) only 4305 N and thus underestimates the experimental load similarly as observed for the thinner 4-ply specimens.

It is thus summarized that the analytical expression for the face sheet rupture force F_{rup} from equation (3.38) generally agrees with the experimental results while the expression for the core shear failure force F_{cs} from equation (3.40) underestimates the observed failure force noticeably. The analytical equations (3.23) and (3.29) that describe damage onset by core crushing and face sheet delaminations may not be confirmed as the influence of oscillations and dynamic effects may override the results. Experimental confirmation of the damage onset criteria should instead focus on indentation testing of specimens with different core thicknesses with subsequent sectioning of the test specimens for confirmation of the discussed failure modes.

Investigation of the full impact response is generally very time consuming for more than only a few specimens. Thus the peak impactor force as the most prominent value is shown in Figure 4.16 while maximum impactor displacement is provided in appendix C.1. The peak force depends mainly on face sheet thickness and applied impact energy. Influence of the face sheet thickness is dominated by the fact that the peak force is limited by face sheet rupture. In case of the 4-ply specimens (diamond symbols) the peak force does not exceed about 6000 N independently of the applied impact energy. For the 6 and 8-ply specimens (square and circle symbols) this increases to about 9000 N and 12000 N respectively. Thus adding two NCF plies to the face sheets increases the peak load by about 3000 N. This correlates with the results of the 2-ply specimens (triangular symbols) which do not exceed about 3000 N. This is also in agreement with the analytical model of equation (3.38) which predicts face sheet rupture loads of 2731 N, 5462 N and 8192 N for 2, 4 and 6-ply face sheets.

In case of impact energy a threshold is required for face sheet rupture. During the impact, the sandwich plate picks up a significant amount of energy through elastic deformation and minor damages until the face sheet rupture force is reached. This initiates for the 2-ply specimens already at 20 J impact energy, while the 4-ply specimens require 35 J and the 6 and 8-ply specimens require even 50 J. These values are only true for the described boundary conditions as the amount of elastic energy stored due to bending increases significantly with larger specimens. Larger sandwich plates are capable of storing more energy via elastic deformation as reported by Tomblin et al. [Tom04].

Sandwich bending stiffness, which is determined by the core and face sheet thicknesses, may not be used as a parameter for distinguishing the impact response alone. The bending stiffness can be altered by both increasing the core or the face sheet thickness which each have contradicting influences on the impact response and resulting failure mode.

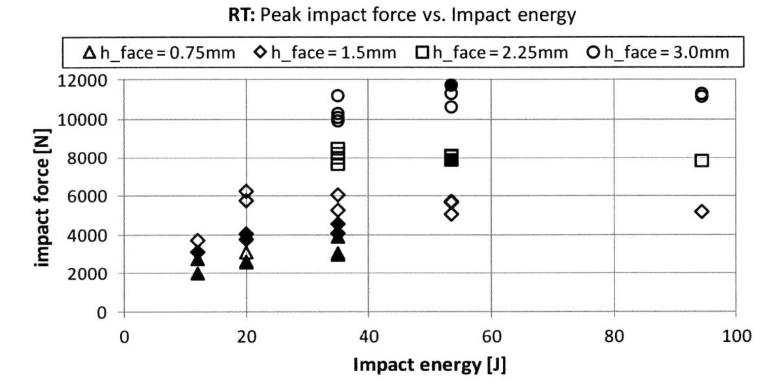

Figure 4.16 – RT impact tests: Peak impact force vs. impact energy (left); empty symbols: Face sheet rupture, filled symbols: Core shear failure

4.3 Test results at low temperatures

In the test series Frost impacts were conducted at low temperatures of -55 °C. These tests were performed as a sandwich structure combines different materials with different coefficients of thermal expansion (CTE), leading to thermal stresses inside the structure following a temperature change. The CTE of CFRP is typically about zero which means that the material does not expand or shrink due to temperature changes. The Rohacell PMI foam core has a CTE of $35*10^{-6}$ 1/K [Gut09] and thus expands as temperature increases but shrinks due to decreasing temperatures. The stiffer CFRP however dominates the thermal response of the sandwich and thus forces the CFRP thermal response onto the foam core creating thermal stresses. As manufacturing takes place at 180 °C curing temperature, the cool down to room temperature introduces already tensile stresses in the foam core. These stresses add to chemical shrinkage due to resin curing in the face sheets, but partially relax due to viscoelastic effects and moisture expansion. Quantifying residual thermal stresses is a complex process as discussed by Brauner et al. [Bra11] and John et al. [Joh11].

The highest thermal stress level is thus present at the lowest operating temperature. Aircraft structures typically have to be qualified up to -55 °C as the lowest service temperature which was thus chosen for the tests. Application of the tool drop impact scenario at this temperature remains questionable as -55 °C occurs only in-flight. The tool drop was however chosen to make the Frost test results comparable to the room temperature results.

The damage types observed include face sheet and core damage similar to the room temperature tests, but also vertical foam cracks as a new damage type. Figure 4.17 shows a vertical foam crack below the point of impact and the related NDI image. The influence of low temperatures on the sandwich failure mode can only be estimated based on the available test results due to the relatively thick core of the tested configurations. None of these were subject to core shear failure when tested at room temperature. NDI images of all Frost specimens are shown in Figure 4.18 in addition with dent depth and applied impact energy. One specimen (Frost_25.7-2.25_P3) experienced core shear failure at an impact energy of 50 J, which is in contrast to the RT tests. At room temperature the specimen RT_H2b_P2, which has the same sandwich configuration as Frost_25.7-2.25_P3, failed instead by face sheet rupture.

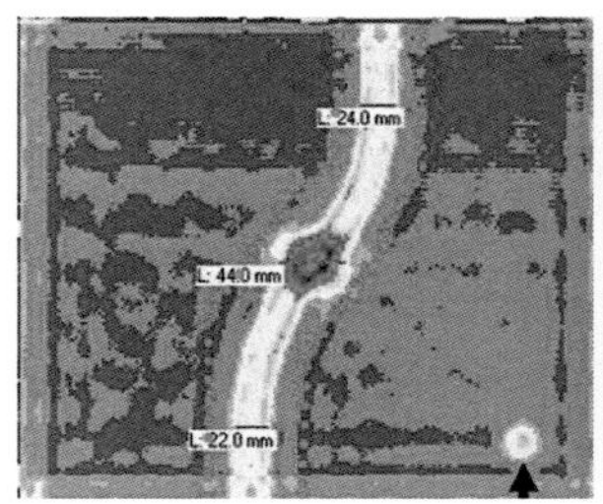

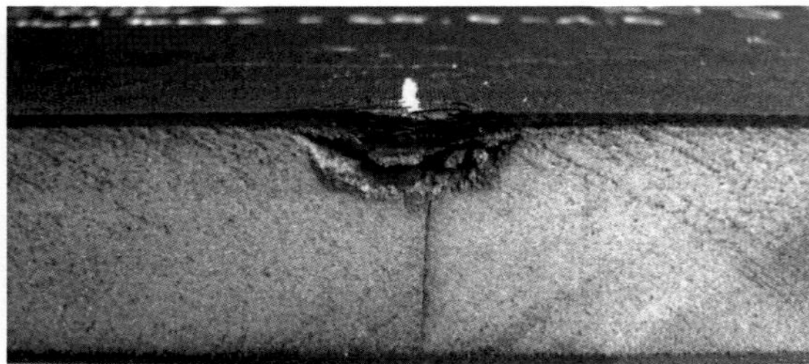

Figure 4.17 – specimen Frost_35.5-1.5_P3: NDI image (left, [Bie11]) and sectioned specimen (right); vertical foam core cracks after a 50 J impact at -55 °C

Vertical foam cracks were observed in three out of four sandwich configurations when tested with 35 J or 50 J impacts. In all cases the cracks are oriented perpendicular to the face sheet principal fiber orientation (0° direction). The cracks pass through the impact point while there are no start and end points clearly visible. All cracks extend through the entire length of the sandwich plate but stop just short of the end of the sandwich plate.

Typically low temperatures cause polymers including polymer foams to become stiffer but also embrittle. This may explain larger damage areas in the foam core but not necessarily vertical foam cracks or its extension outside of the area affected by the impact load. One noticeable issue is however that the cracks align always perpendicular to the principal direction of the fiber orientation of the face sheets. The CTE mismatch causes thermal tensile stresses in the core. As the CTE of the face sheets is anisotropic, the largest CTE mismatch originates between the core and the face sheet in the direction of the principal fiber orientation. Thus a reasonable explanation for the vertical foam core cracks is that the impact locally induces a core crack that is then driven by thermal stresses.

The exact mechanism of crack initiation is however unclear. One hypothesis is that the core crushing process and the resulting damage start local cracks that act as a crack starting notch while crack growth is then driven by tensile thermal stresses in the foam core. A second hypothesis is that a superposition of sandwich bending and thermal tensile stresses in the core generate a sufficiently large crack. For clarification an additional -55 °C impact test was performed with 50 J impact energy and modified boundary conditions using specimen RT_H2c_P3, which was initially envisaged for RT tests and is identical to specimen Frost_35.5-2.25_P3. The sandwich plate was now placed on a continuous stiff support eliminating bending of the sandwich plate as shown in Figure 4.19. The NDI image of the specimen reveals a vertical foam crack comparable to those previously observed.

In consequence the observed vertical foam cracks are primarily understood as a product of the low temperature and the sandwich failure mode face sheet rupture. This failure mode generates sufficient foam crushing to start a foam crack, which is then driven by thermal stresses. This agrees with the observation that only specimens with a dent depth of at least 2.5 mm experienced vertical foam cracks. The dent depth may thus be used a measure of the severity of impact damage with respect to starting vertical foam core cracks.

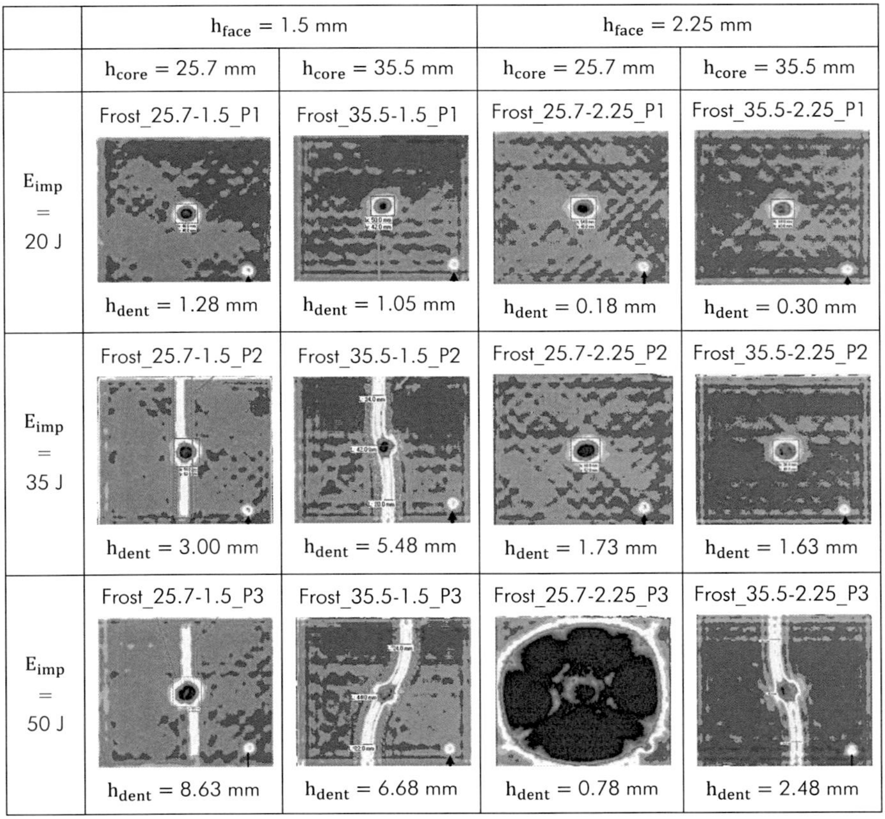

	h_{face} = 1.5 mm		h_{face} = 2.25 mm	
	h_{core} = 25.7 mm	h_{core} = 35.5 mm	h_{core} = 25.7 mm	h_{core} = 35.5 mm
E_{imp} = 20 J	Frost_25.7-1.5_P1 h_{dent} = 1.28 mm	Frost_35.5-1.5_P1 h_{dent} = 1.05 mm	Frost_25.7-2.25_P1 h_{dent} = 0.18 mm	Frost_35.5-2.25_P1 h_{dent} = 0.30 mm
E_{imp} = 35 J	Frost_25.7-1.5_P2 h_{dent} = 3.00 mm	Frost_35.5-1.5_P2 h_{dent} = 5.48 mm	Frost_25.7-2.25_P2 h_{dent} = 1.73 mm	Frost_35.5-2.25_P2 h_{dent} = 1.63 mm
E_{imp} = 50 J	Frost_25.7-1.5_P3 h_{dent} = 8.63 mm	Frost_35.5-1.5_P3 h_{dent} = 6.68 mm	Frost_25.7-2.25_P3 h_{dent} = 0.78 mm	Frost_35.5-2.25_P3 h_{dent} = 2.48 mm

Figure 4.18 – Frost impact tests: NDI results of all specimens with clamped boundary conditions [Bie11]

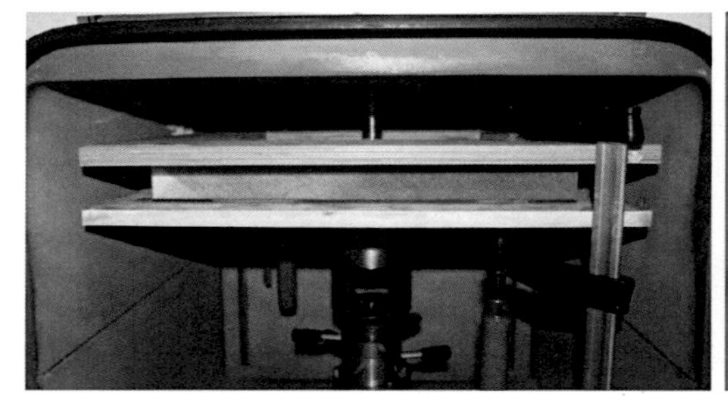
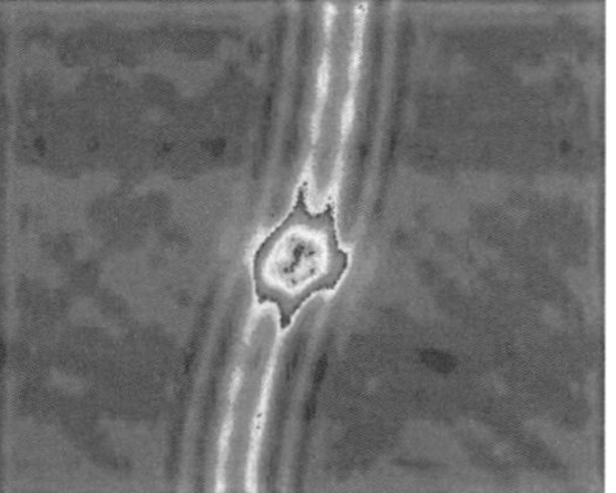

Figure 4.19 – impact testing of specimen RT_H2c_P3 at -55 °C: Test setup (left) [Rin10], NDI image following a 50 J impact (right) [Fri11]

A core shear failure was observed in case of specimen Frost_25.7-2.25_P3 which was impacted with 50 J. Impact testing of the same configuration RT_H2b with 50 J at room temperature was performed with specimen RT_H2b_P2 and did instead lead to face sheet rupture. Increasing the impact energy to 90 J as performed with specimen RT_H2b_P3 did not alter this result. This result may be affected by statistical scatter, but Figure 4.20 compares the sectioned test specimens Frost_25.7-2.25_P3 and RT_H2a_P2 both subject to core shear failure. Configuration RT_H2a is identical to RT_H2b and Frost_25.7-2.25, but has a thinner foam core of only 16.3 mm compared to 25.7 mm for the other specimens. It failed by core shear failure in the RT test series.

Figure 4.20 – appearance of core shear failure: Ductile failure with a mixture of local plastic deformation and stepwise cracking across multiple foam cells in specimen RT_H2a_P2 at room temperature (top); largely brittle fracture in specimen Frost_25.7-2.25_P3 at -55 °C (bottom)

When comparing the different appearances of core shear failure it becomes clear that the failure during -55 °C tests is more brittle as a single crack extends through the core. In contrast the shear failure at room temperature is characterized by a mixture of local plastic deformation of foam cell walls and stepwise cracking across multiple cells which also leads to lower visibility.

It is thus summarized that low temperatures lead to two noticeable changes of the impact response. First vertical foam cracks appear in specimens subject to face sheet rupture. These cracks are initiated by either the core crushing process or the resulting foam damage and then driven by thermal stresses. Secondly a higher sensitivity to core shear failure is noticed. Despite limited statistical relevance the appearance of the core shear failure permits the conclusion that thermal embrittlement is a factor contributing to this and in light of the vertical foam cracks detrimental to the overall impact response.

The impact damage is characterized by damage mode, size and dent depth as a measure of its visibility. Figure 4.21 and Figure 4.22 display the planar damage diameter and residual indentation from series RT and Frost. Only specimens without shear cracks and core thicknesses of 25.7 or 35.5 mm are included in order not to obscure the results. These show that temperatures of -55 °C have a slightly detrimental effect as the typical damage size increases from approx. 40 mm for a 35 J impact at room temperature up to 60 mm for a 35 J impact at -55 °C. It must be noted that at least some damage sizes may be affected by the appearance of vertical foam cracks, as damage size was measured using the NDI images. These do not allow a clear distinction of vertical cracks and core crushing damage. Dent depth finally ranges for a 35 J impact from a little above 1 mm for the specimens with 2.25 mm face sheet thickness to about 5.5 mm for the specimens with 1.5 mm face sheets without a noticeable temperature influence as shown in Figure 4.22.

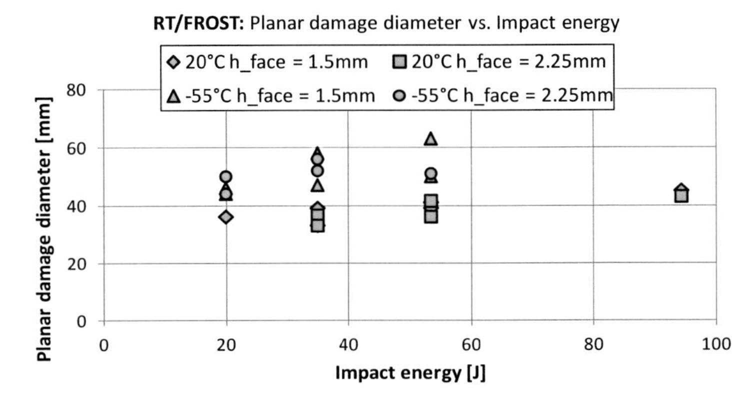

Figure 4.21 – RT and Frost impacts: Damage size vs. impact energy;
only specimens with core thicknesses of 25.7 mm and 35.5 mm without shear failure shown
(blue symbols: -55 °C tests; green symbols: 20 °C tests)

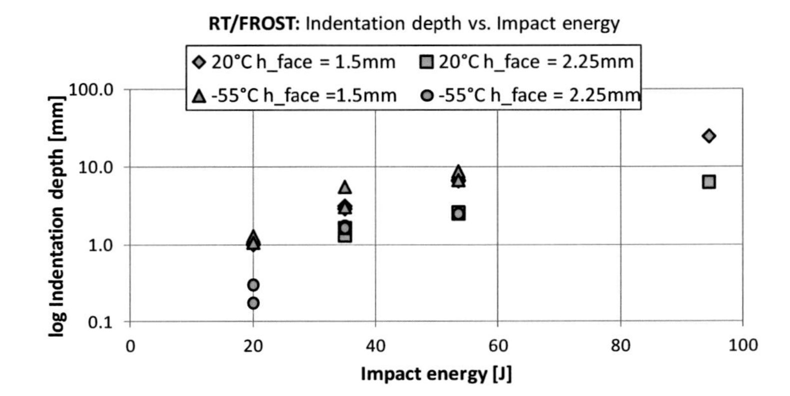

Figure 4.22 – RT and Frost impacts: Residual indentation depth vs. impact energy; only specimens with core thicknesses of 25.7 mm and 35.5 mm without shear failure shown (blue symbols: -55 °C tests; green symbols: 20 °C tests)

The dynamic impact response of the sandwich is characterized by the force history and force vs. displacement data of the impactor as shown in Figure 4.23 and Figure 4.24 for 50 J impacts on RT and Frost specimens with skin thicknesses of 1.5 mm and 2.25 mm.

Comparing the different curves one difference was initially noted when comparing the plateau value. This is the force level, that is held more or less steady, after the impactor ruptured the skin and continuously penetrates deeper into the sandwich. This plateau value is for the 1.5 mm face sheet specimens tested at -55 °C about 1000 N lower than for the same specimens when tested at room temperature. Consequently max. impactor displacement increases similarly.

This trend does however not visualize when comparing the 2.25 mm test specimens. Here the lower impactor force of specimen Frost_25.7-2.25_P3 is related to core shear failure. Once the shear crack starts, the specimen loses a great portion of its bending stiffness which explains the lower plateau force and larger displacement compared to the identical specimen tested at room temperature. The specimen Frost_35.5-2.25_P3 does however not show a noticeable lower plateau value than its counterpart at room temperature and thus not support the observation from the 1.5 mm face sheet specimens.

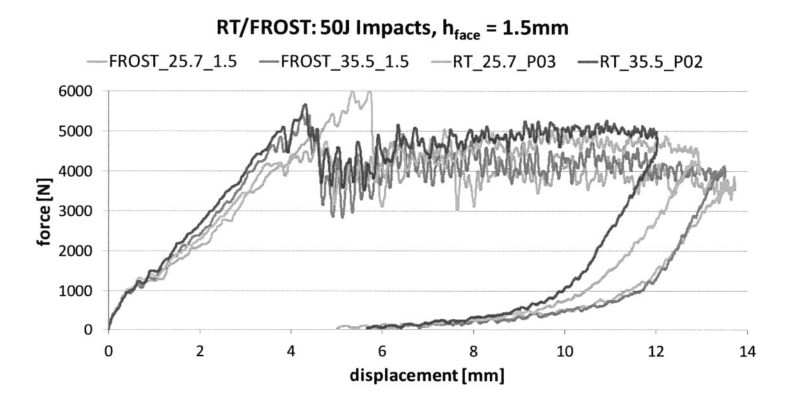

Figure 4.23 – RT and Frost specimens: Force vs. displacement of 50 J impacts on specimens with 1.5 mm face sheets

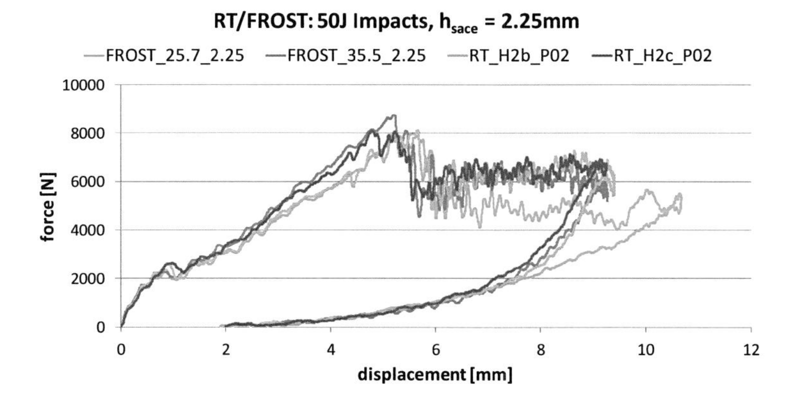

Figure 4.24 – RT and Frost specimens: Force vs. displacement of 50 J impacts on specimens with 2.25 mm face sheets

4.4 Classification of damage types

Sandwich impact damage may now be classified into four categories related to the failure modes. The selected categories and corresponding impact damages are depicted in Figure 4.25 schematically and Figure 4.26 using pictures of sectioned specimens: (1) Indentation damage leads to (a) face sheet rupture with significant foam core crushing and (b) crushing induced local shear cracks at high impact energies. (2) Core shear failure leads to core shear damage and cracks while (3) residual thermal stresses in the core lead to vertical foam core cracks. This thermal damage requires face sheet rupture and sufficient core crushing as a prerequisite. An additional category (0) is introduced for local face sheet and core damage that do not relate to any specific sandwich failure mode.

Type 0 local face sheet damage and core crushing are the result of relatively low energy impacts that stay below the energy threshold for either of the failure modes core shear failure or face sheet rupture. Impact damage is thus limited to local delaminations and matrix cracks in the face sheet and core crushing below the point of impact. Core crushing occurs once the compressive foam strength is locally superseded as described in equation (3.23). The size of the crushed core zone grows with impact energy and face sheet thickness. Core crushing has a low visibility until face sheet rupture occurs as the damaged face sheet is typically subject to spring back. Face sheet delaminations occur due to local bending of the impacted face sheet and may be described using equations (3.28) and (3.29).

Type 1 indentation damage is the result of the failure mode face sheet rupture and splits into (a) foam core crushing and (b) crushing induced core shear cracks. In addition to the damages described for type 0, face sheet damage now includes also fiber cracking. Damage size is typically somewhat larger than for type 0, but its severity and depth increase due to rupture of the face sheet. Also the impactor penetrates deeper into the sandwich thus creating more severe core crushing.

Type 1 (b) crushing induced core shear cracks can be observed at very high impact energies relative to the sandwich configuration and must not be confused with damage type 2. In this special case the impactor ruptures the face sheet, penetrates deeply into the foam core and creates at its flanks a zone with crushed core material. As it now approaches the rear face sheet the crushed core debris pushes against the rear face sheet and thus acts as a crack starter. Debris from the ruptured face sheet may contribute to this. Once a crack started it grows roughly at 45° relative to the sandwich plane until it reaches the rear face sheet where it locally continues. The planar damage size of damage type 1(b) is thus larger than that of type 1(a), but still significantly smaller than the extent of damage type 2 as shown in Figure 4.10 for specimens RT_10.0_P2 and P3.

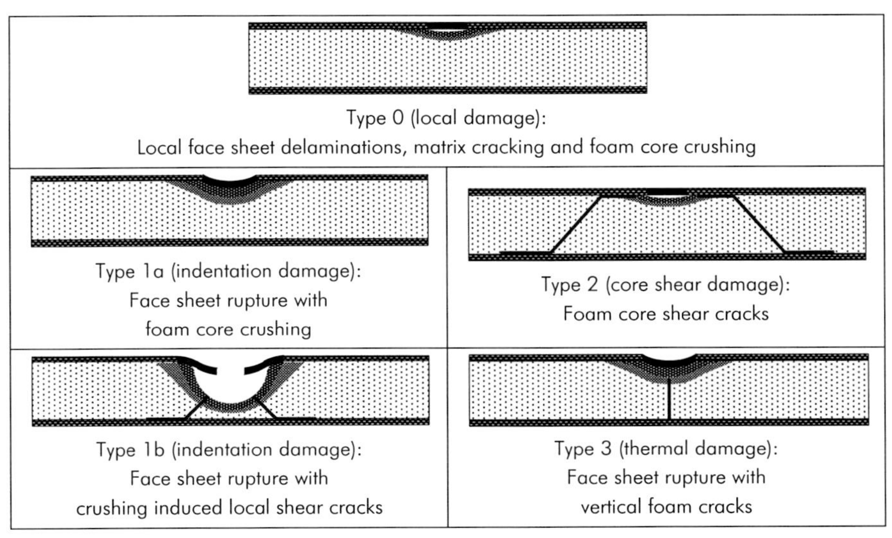

Figure 4.25 – schematic description of observed damage types after low velocity impact

Type 2 foam core shear failure will occur, if core shear stresses supersede the shear strength and fracture toughness of the foam. The shear load emerges in a circular area around the impact due to bending of the sandwich plate and can be determined with equation (3.39). Depending on the temperature the shear failure may be more ductile or brittle as shown previously in Figure 4.20. The observed damage consists of local core crushing and a foam crack that spans from the upper to the lower face sheet at 45° to 60° relative to the sandwich plane. The damage has a circular shape and continues further along the rear face sheet interface creating a hat shaped contour. Damage size is significantly larger than that of damage type 1.

Type 3 foam core vertical cracks are a different kind of damage and related to thermal stresses in the foam core. These stresses originate due to a CTE mismatch of the face sheets and the polymer foam core. In the investigated case the face sheets have a CTE close to zero while the foam has a positive CTE typical for polymeric materials. Thus the highest stresses emerge at low temperatures with the lowest in service temperature for aircraft structures being -55 °C. Impacts at this temperature hit an already preloaded structure. Severe local core crushing now acts as a crack starter notch due to the brittleness of the Rohacell RIST foam. The vertical foam crack then orients perpendicular to the largest residual tensile stress, which is in the investigated cases the direction of the principal fiber orientation of the face sheets. Consequently this damage type is not impact damage but initiated by the impact. As thermal stresses in the foam core are the driver behind it, this damage type is also referred to as thermal damage or impact induced thermal cracks.

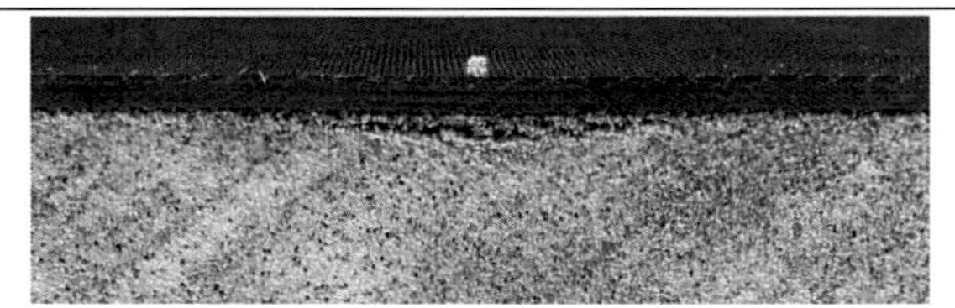

Type 0 (local damage):
Local face sheet delaminations, matrix cracking and foam core crushing

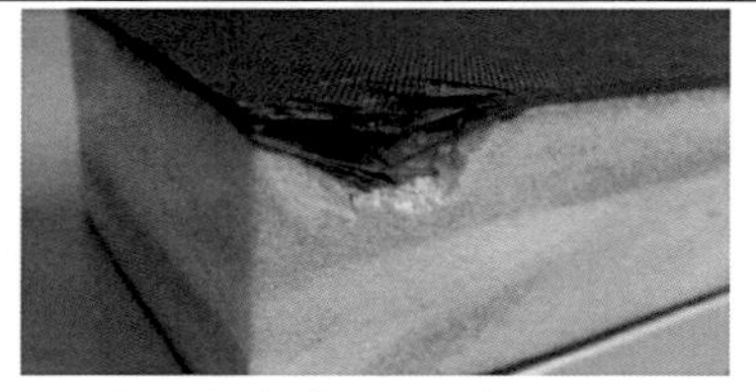

Type 1a (indentation damage):
Face sheet rupture with core crushing

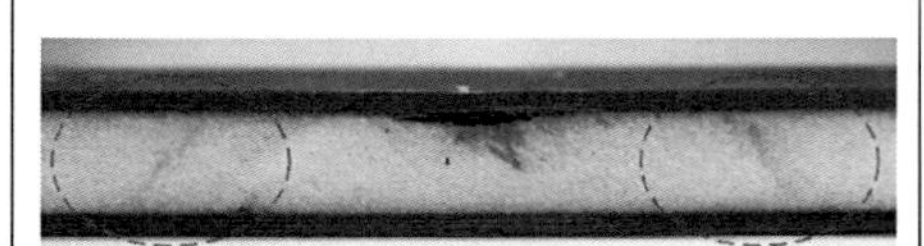

Type 2 (core shear damage):
Foam core shear cracking

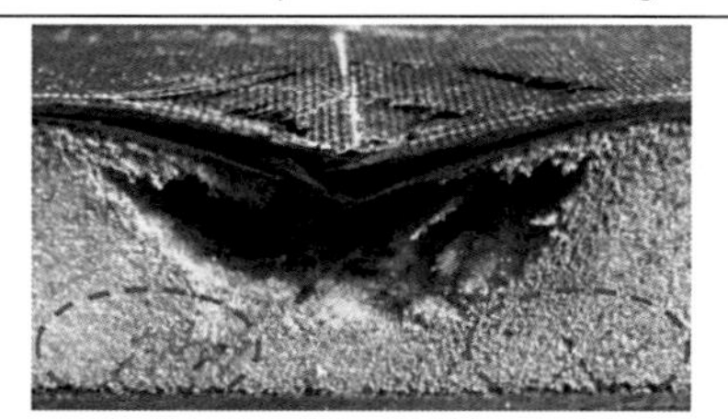

Type 1b (indentation damage):
Face sheet rupture with
crushing induced shear cracks

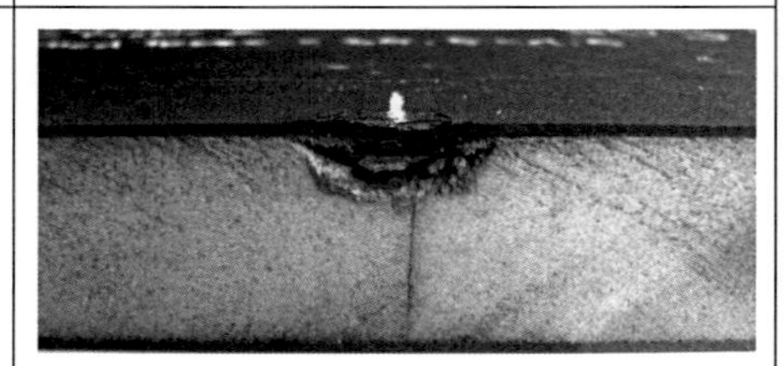

Type 3 (thermal damage):
Face sheet rupture with
vertical foam cracks

Figure 4.26 – observed damage types after low velocity impact

4.5 Comparison with analytical model

A failure mode map has been created for low velocity impact loading of composite foam core sandwich plates in section 3.5. This map is based on the analytical failure criteria for face sheet rupture and core shear failure. Figure 4.27 now shows a comparison of the failure mode map with the experimentally observed impact damage of all RT test specimens.

The analytical shear failure criterion of equation (3.42) and the test results are overall in accordance with one another. The failure criterion seams to overestimate the danger of shear cracks slightly as already discussed in sections 4.2 when presenting the force history and displacement data of selected RT tests. The applied material properties used for creation of the failure mode map were taken from appendix B and are thus based on test results and reverse engineered properties, which may contribute to the overestimation of the shear crack probability. The parameter $\bar{a}^2 \leq 0.8$, which describes the effect of membrane stresses on the core shear load, was set to the maximum recommended value of $\bar{a}^2 = 0.8$ for a conservative calculation of mode transition.

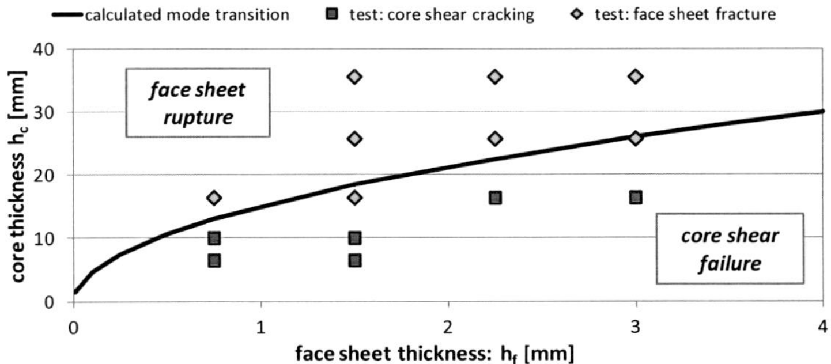

Figure 4.27 – impact failure mode map for low velocity impact of foam core composite sandwich plates with analytical solution and RT test results

4.6 Chapter summary

Two parametric low velocity impact test series were conducted on CFRP sandwich specimens made of a Rohacell 71RIST foam core. Impact energies in the test series ranged from 12 J to 90 J. The first test series (RT) investigated the impact response at room temperature while the second test series (Frost) investigated the same at low temperatures of -55°C. Test results confirmed the occurrence of two governing sandwich impact failure modes: Face sheet rupture and core shear failure.

Core shear failure leads to low visibility and limited damage of the impacted face sheet but initiates large core damage of sizes up to 250 mm diameter or more. Face sheet rupture instead leads to a better visibility as face sheet damage becomes more severe and residual dent depth increases. Core damage is limited to local core crushing around the point of impact. An interaction of core shear failure and face sheet rupture was observed for specimens with relatively high impact energies that fail at lower impact energies by core shear failure only. Thus if in reverse the threshold impact energy of face sheet rupture is lower than the energy required for face sheet rupture, no core shear failure will follow. If however the threshold impact energy of core shear failure is lower and thus it occurs first, higher impact energies may still lead to face sheet rupture due to continuous membrane loading of the face sheet and sandwich.

The impact energy required to cause either face sheet rupture or core shear failure varied significantly depending on the sandwich configuration. Lighter panels with a core thickness of less than 10 mm and face sheet thicknesses of 0.75 mm or 1.5 mm were subject to core shear failure at only 12 J while specimens with a core thickness of 35.5 mm and face sheet thicknesses of 2.25 mm or 3 mm required 50 J impact energy to fail by face sheet rupture.

Low temperature impact tests at -55 °C revealed additionally vertical foam core cracks. This damage type was not observed at room temperature and emerged only together with face sheet rupture. The appearance and orientation of vertical foam cracks point to the conclusion that the cracks are driven by thermal strains but triggered by significant core crushing damage following face sheet rupture.

The resulting impact damages were thus finally categorized in agreement with the failure modes:

- Type 0 local impact damage describes local face sheet delaminations, matrix cracking and core crushing. This damage type is limited to impacts with less energy than required for either of the governing sandwich impact failure modes to occur.
- Type 1 indentation damage is based on the failure mode face sheet rupture and includes in addition to the same (a) core crushing and (b) severe core crushing combined with local shear cracks.
- Type 2 core shear damage is categorized by shear cracks in the foam core which emerge due to the failure mode core shear failure at an angle of 45° to 60° relative to the face sheet. The shear cracks have a hat shaped contour and continue as a rear face sheet debonding.
- Type 3 thermal damage describes impact induced vertical foam cracks.

Finally the test results were compared with the previously developed damage mode map and show that the analytical model slightly overestimates the criticality of shear failure but otherwise provides a good prediction of the failure mode. Furthermore it becomes clear that the low visibility failure mode core shear failure (damage type 2) can be avoided by designing sandwich structures with sufficient core thickness.

5 Numerical simulation of impact on CFRP foam core sandwich structures

5.1 General aspects

Numerical simulation using the finite element method (FEM) has become a standard tool in the development of industrial products. Its application reduces the development effort of new products significantly as it avoids excessive manufacturing and testing of prototypes. In the aerospace industry numerical simulation is performed in the context of the building block approach (BBA) as described previously in section 3.1. Test and simulations are performed in parallel with stepwise increasing complexity. Each test confirms a particular detail on one level, which leads up to a single full scale test. If the simulation results are confirmed by the test results, it will be possible to use the same simulation approach to validate further improvements as long as the design principle of the structure does not alter and thus the governing failure modes remain identical.

As impact is a short time dynamic problem an explicit finite element (FE) code is desirable to analyze the impact response. In this work the explicit FE code LS-DYNA is used due to previous experience and a large available database of predefined material models for composite materials and polymeric foam cores. Similar explicit finite element codes such as e.g. PAM-CRASH or Abaqus/Explicit are in principal capable of the same.

5.1.1 Basic equations and explicit time integration scheme

The expression implicit and explicit finite element code is related to the integration scheme used for solving the governing differential equations in the time domain. Implicit solvers are commonly applied for static and quasi-static problems. Here relatively large time steps are applied for solving the problem as the implicit time integration scheme can solve these stable as long as only limited nonlinear behavior is present. Short time dynamic problems such as wave propagation, crash and impact are typically simulated using explicit solvers. Explicit solvers can solve the problem only in short time steps but can cope well with a high degree of nonlinear behavior. Due to the short time steps they deliver a high resolution in the time domain which is typically asked for when analyzing short time dynamic problems.

The FEM is based on the idea that structures and loads are discretized in geometry and time. Geometrical discretization is performed using finite elements that divide a complex geometry into simple geometries such as bars, triangles, rectangles or their equivalent volumetric shapes. These can now be analyzed using simple basic functions with e.g. linear or quadratic behavior. If the discretization is sufficiently fine and thus enough elements are used to discretize the structure, it will be possible to describe a complex stress distribution with simple base functions within a geometrically complex structure. The same approach is applied to describe the load history. Hence the time is then discretized in time steps.

The basic equation of motion of a harmonic oscillator is described by [Ste09, Lst06]

$$m\,\ddot{u}(t) + d\,\dot{u}(t) + k\,u(t) = F(t)\ . \tag{5.1}$$

Here m is the mass, d the damping coefficient, k the spring stiffness, $u(t)$ the displacement of the oscillator and $F(t)$ the applied external force history. The same equation can also be set up for a multi-mass harmonic oscillator where $[m]$, $[d]$ and $[k]$ become the equivalent matrices for mass, damping and stiffness while the displacement $\mathbf{u}(t)$ becomes a vector. Similarly the external force $\mathbf{F}(t)$ is a vector function, too. Thus the equation of motion for a multi-mass harmonic oscillator becomes [Hei08]

$$[m]\,\ddot{\mathbf{u}}(t) + [d]\,\dot{\mathbf{u}}(t) + [k]\,\mathbf{u}(t) = \mathbf{F}(t) \tag{5.2}$$

The dynamic behavior of a mechanical system discretized by multiple finite elements can be described by such an equation. Solving this equation of a multi-mass oscillating system is thus a central part of the FEM. As already mentioned the time t is also discretized in dynamic problems. Thus equation (5.2) can be set up both at an initial time t_n and the time $t_{n+1} = t_n + \Delta t$, which is one time step Δt later. Equation (5.2) now becomes

$$[m_n]\,\ddot{\mathbf{u}}_n + [d_n]\,\dot{\mathbf{u}}_n + [k_n]\,\mathbf{u}_n = \mathbf{F}_n \qquad \text{and} \tag{5.3}$$

$$[m_{n+1}]\,\ddot{\mathbf{u}}_{n+1} + [d_{n+1}]\,\dot{\mathbf{u}}_{n+1} + [k_{n+1}]\,\mathbf{u}_{n+1} = \mathbf{F}_{n+1}\ . \tag{5.4}$$

In practice two principal ways are used to determine the unknown displacement $\mathbf{u}_{n+1}$ and its derivatives. The first way is known as the implicit time integration and sets up equation (5.4) and solves for equilibrium using a numerical solution scheme such as e.g. the Newmark method [New59]. Typically it is assumed that the matrices $[m]$, $[d]$ and $[k]$ are constant and the applied external load $\mathbf{F}_{n+1}$ is known. As solving equation (5.4) for the displacement $\mathbf{u}_{n+1}$ is mathematically very time consuming and requires – due to reordering of its elements – inversion of the stiffness matrix $[k]$, this approach is called an implicit solution and is computationally rather expensive. If nonlinear behavior such as material plasticity or contact appear, this will become more complex as particularly $[k]$ is not anymore constant. Equation (5.4) now has to be solved iteratively.

The implicit method works well as long as the excitation force varies moderately and nonlinear behavior remains modest allowing application of relatively large iterative time steps. Severe nonlinear behavior may lead to non convergence of the solution and thus make solving equation (5.4) impossible.

The second way of determining the unknown displacement $\mathbf{u}_{n+1}$ is the explicit time integration which is typically based on finite difference methods for numerically solving the equation of motion. Here equation (5.3) is used as the starting point at the initial time t_n and then extrapolated by one time step Δt to the new time t_{n+1}. Assuming a linear base function and thus linear behavior during the time step as shown in Figure 5.1, the velocity $\dot{\mathbf{u}}$ at different times can be described by the relationships

$$\dot{\mathbf{u}}_{n+1/2} = \frac{1}{\Delta t}(\mathbf{u}_{n+1} - \mathbf{u}_n) \quad \text{and}$$
$$\dot{\mathbf{u}}_{n-1/2} = \frac{1}{\Delta t}(\mathbf{u}_n - \mathbf{u}_{n-1}) \, . \tag{5.5}$$

The velocity $\dot{\mathbf{u}}_n$ and acceleration $\ddot{\mathbf{u}}_n$ at the time $\mathbf{t}_n$ now become

$$\dot{\mathbf{u}}_n = \frac{\dot{\mathbf{u}}_{n+1/2} + \dot{\mathbf{u}}_{n-1/2}}{2} = \frac{1}{2\,\Delta t}(\mathbf{u}_{n+1} - \mathbf{u}_{n-1}) \quad \text{and} \tag{5.6}$$

$$\ddot{\mathbf{u}}_n = \frac{1}{\Delta t}\left(\dot{\mathbf{u}}_{n+1/2} - \dot{\mathbf{u}}_{n-1/2}\right) = \frac{1}{\Delta t^2}(\mathbf{u}_{n+1} - 2\,\mathbf{u}_n + \mathbf{u}_{n-1}) \, . \tag{5.7}$$

Inserting equations (5.6) and (5.7) into the equation of motion (5.3) yields the following equation (5.8), which can be solved directly for the unknown displacement $\mathbf{u}_{n+1}$. Because of the direct character of the solution this way of solving the dynamic problem is referred to as explicit time integration scheme:

$$\left(\frac{1}{\Delta t^2}[m_n] + \frac{1}{2\,\Delta t}[d_n]\right)\mathbf{u}_{n+1} =$$
$$\mathbf{F}_n - \left([k_n] - \frac{2}{\Delta t^2}[m_n]\right)\mathbf{u}_n - \left(\frac{1}{\Delta t^2}[m_n] - \frac{1}{2\,\Delta t}[d_n]\right)\mathbf{u}_{n-1} \, . \tag{5.8}$$

If the matrices $[m]$ and $[d]$ are diagonal matrices, it will be possible to solve equation (5.8) with limited numerical effort as no matrix inversions are necessary. It has to be noted however that knowledge of the results of the previous two time steps is required as both displacements $\mathbf{u}_n$ and $\mathbf{u}_{n-1}$ are used in equation (5.8). Thus the explicit integration is a two-step method and requires a separate solution for the initial integration step.

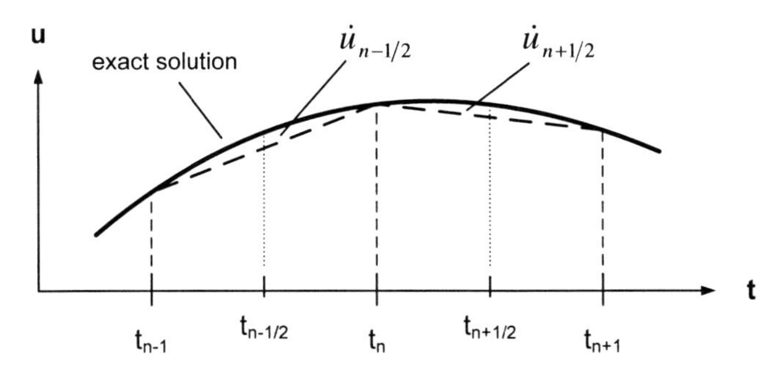

Figure 5.1 – finite difference method used for explicit time integration [Hei08]

Selection of the time step Δt has a mayor influence on the effort required for solving the problem and the quality of the results. A necessary condition for mathematical convergence of the finite difference method is the Courant-Friedrichs-Lewy condition [Cou28]. This condition states that physically the time step Δt must be smaller than the time required for a sound wave to pass through the smallest element in the mesh. Based on the Courant-Friedrichs-Lewy condition a critical time step Δt_{crit} can be defined that must not be exceeded. The critical time step in LS-DYNA is defined as [Lst06, Ste09]

$$\Delta t \leq \Delta t_{crit} = \frac{L_{char}}{c}\,. \tag{5.9}$$

Here L_{char} is the characteristic length of the element and c the speed of sound in the material. The speed of sound is defined in LS-DYNA depending on the element type. For shell elements it becomes [Lst06,Ste09]

$$c = \sqrt{\frac{E}{\rho(1-\nu^2)}}\,. \tag{5.10}$$

The critical time step Δt_{crit} thus depends primarily on the selected element type and size as well as the material properties stiffness and density. In anisotropic materials the speed of sound varies depending on the material orientation. Thus L_{char} must be checked in each direction of the element as the shortest element edge or height may be due to a lower speed of sound in this direction not necessarily the critical dimension.

As element length is typically selected based on the level of accuracy required and the elastic stiffness E is an important material parameter that must not be altered, only the density may be used as a parameter to increase the critical time step and thus reduce calculation time. This technique is referred to as mass scaling and commonly applied to problems, where the exact mass or a local mass change are not of interest and do not alter the results significantly.

Following determination of the displacements, strains and strain rates can be calculated directly while stresses are determined using the constitutive law of the applied material.

5.1.2 Modeling approach

Sandwich structures can be modeled by finite elements using different discretizations. The level of discretization is determined based on the accuracy required for the individual task. Figure 5.2 summarizes relevant discretizations of composite sandwich structures. Here the sandwich components core and skin are used for distinguishing. The interface as the third sandwich component may either be omitted by using a fixed connection between skin and core or by adding a defined connection with a failure option.

The coarsest discretization of a sandwich structure is the homogenized sandwich which is based on a single layered shell. Here the properties of each material layer are added together using laminated shell theory. The selected approach must include shear deformations as sandwich structures typically have a thick but low shear stiffness core which contributes significantly to the total deflection. This approach is typically capable to describe the sandwich stiffness correctly and may thus be applied in areas where this is the major concern. If however stresses are of interest, a finer approach will likely be required.

The sandwich can be split into separate elements for face sheet and core in order to improve the accuracy of the simulation. This is typically performed by using one layered shell element for each face sheet and volume elements for the core. Alternatively layered solid elements or mixed shell-solid formulations can be used that account for stresses in the thickness direction of the face sheet. The volume elements of the core are now capable of describing the full three dimensional (3D) stress state of the core and thus allow a correct determination of its stress state. Due to separate core and face sheet elements it is also possible to determine interface stresses for analyzing e.g. face sheet debonding.

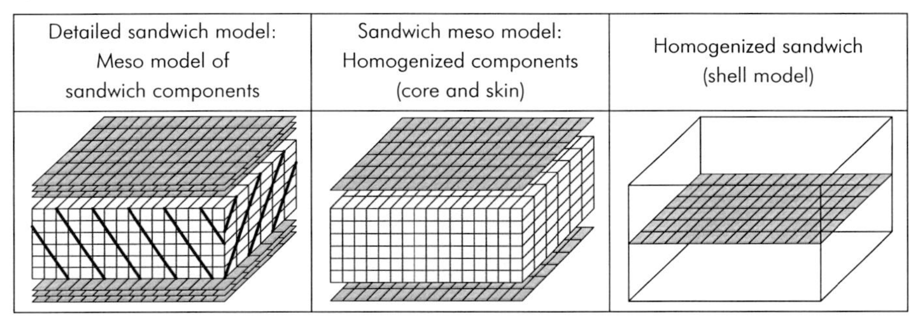

Figure 5.2 – discretization levels of composite sandwich structures

In the case of inhomogeneous cores such as honeycombs or folded cores, the core structure may also be described using shell elements representing individual cell walls, which will be referred to as a meso core model [Hei07]. This idea can be extended to foam core structures with discrete core reinforcements such as pins as these may be included in the model using e.g. bar elements. Furthermore the skins may be split into separate shell elements to investigate the interlaminar behavior of the composite laminate to track delaminations and improve the model accuracy. The different layers of the face sheet may now e connected using a cohesive zone model embedded either within special purpose elements or a more general contact formulation.

As this investigation concentrates on the impact response and resulting failure modes, stresses in the sandwich constituents have to be known. Shear deflection and core shear stresses have to be analyzed, as core shear failure is a major concern. In consequence modeling of the composite sandwich structure is performed using a hybrid sandwich model that composes of a detailed sandwich model and a sandwich meso model. The detailed sandwich model is used at the point of impact and consists of a meso model of the impacted face sheet, a cohesive zone interface model and a homogenized foam core. The impacted face sheet is split into one shell element per ply connected by cohesive zone elements which allows describing delamination as interlaminar failure mode. This model is simplified outside the contact area of the impactor. Here a sandwich meso model is used that consists of one stacked shell element for the face sheets and volume elements for the core. Generally at least five volume elements with linear base functions are used across the thickness of the foam core for correct description of the shear deformation. The interface is modeled using cohesive zone elements in areas were failure may occur and alternatively a tied contact without material failure.

The BBA as described in section 3.1 is used as a guideline for the development of the simulation model. For information Figure 5.3 shows the BBA testing pyramid both generic and as it is applied to the sandwich impact problem. It turned out that some of the planed tests on the element level could not be conducted within the scope of this study and thus had to be replaced by literature results. Tests on the detail level and beyond are not performed as these become component and thus application specific. Analysis verification was thus performed with generic impact tests only.

Coupon tests are used to determine stiffness and strength values of the skin and core materials and allow describing the principal material behavior. For verification of the individual modeling approaches, selected coupon tests are simulated in parallel. The next step are tests on the element level which are in the investigated case typically sandwich specimens. These tests focus on damage mechanisms within the sandwich where different material properties interact with each other. Relevant sandwich tests are bending tests, local indentation of the sandwich, fracture mechanics tests of the sandwich interface and generic impact testing of sandwich specimens. Available test results were limited to the sandwich interface, local indentation and generic impact tests, which were also simulated in parallel.

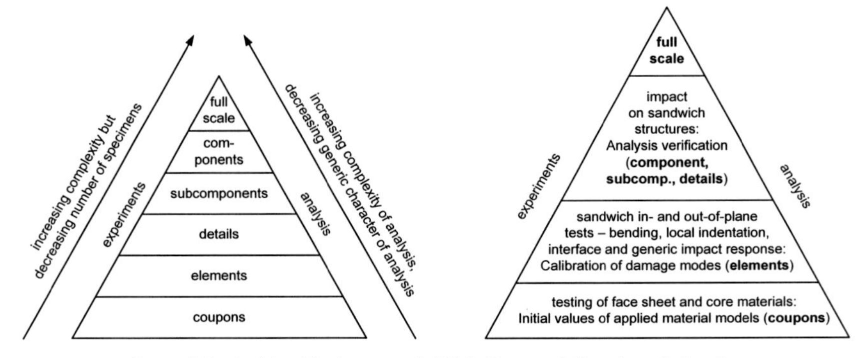

Figure 5.3 – building block approach (BBA): Generic (left) and applied to the impact response of composite sandwich structures (right)

5.1.3 Thermal effects

Besides mechanical loads sandwich structures are also subject to thermal loads as already explained in section 4.3. The implementation of thermal stresses due to a mismatch of different CTE of the face sheet and core can be performed by superposition when assuming linear behavior. Thus a thermal load is applied prior to the mechanical load. The thermal load is based on the cool down of the composite sandwich part from the curing temperature during manufacturing to the temperature present during its later use. The later will be referred to as the working temperature.

This approach is however a simplification as stress relaxation due to viscoelastic processes in the face sheets and the foam core is not accounted for. John et al. [Joh11] describe that a Rohacell RIST foam core may relax to about 70% of its initial stress level when loaded by 1% tensile strain at a temperature of 120 °C for 24 hours. This reduces to 80% of the initial stress level for a temperature of 80 °C and 24 hours of exposure. Also the CTE of most engineering materials, particularly engineering plastics, is temperature dependent.

There may also be chemical shrinkage present due to the curing process of the epoxy resin. Brauner et al. [Bra11] show that in a pin reinforced foam core sandwich structure between 10% and 50% of the combined curing and thermal stresses relax. These numbers however depend on the particular stress state and location within the sandwich being looked at. Also the results are based on theoretical models and were not yet verified by experiments. Thus these numbers have to be used with care and may only provide a rough estimate of the dimension of stress relaxation. Thus selecting the correct curing and thermal stresses for preloading the sandwich structure prior to mechanical loading is difficult.

For determining the sensitivity of sandwich structures to thermal stresses hence different simulations will be performed that aim to clarify the effect of thermal loads. The simulations are based on linear elastic behavior and include cooling down of the sandwich structure from a stress free temperature to a working temperature. Mechanical loading is applied at working temperature only. The thermal load describes three reference situations each based on a 180 °C curing cycle for the sandwich structure. In both cases only limited relaxation is taken into account and assumes a stress-free material state at 170 °C:

- No thermal load: $\mathrm{dT} = 0$ K
- Working temperature 20 °C (reference): $\mathrm{dT} = -150$ K prior to mechanical load
- Working temperature -55 °C: $\mathrm{dT} = -225$ K prior to mechanical load

Additionally an intermediate step of $\mathrm{dT} = -75$ K was used to recognize trends. It is noted that the application of a thermal load in the simulation only includes thermal strains due to a mismatch of CTE. The effect of temperature dependent material parameters is not accounted for. Cool down of polymeric materials typically leads to increased stiffness and strength at the cost of material embrittlement. As thermal loads also have a noticeable effect on the monolithic composite behavior, all simulations in this work where performed using the reference thermal load prior to mechanical loading unless stated differently.

5.2 Modeling of the CFRP face sheets

The applied CFRP is made of a non crimp fabric (NCF) textile semi finished product with three principal fiber orientations of 0°, 45° and -45° and thus of triaxial character. Current practice for modeling laminated CFRP is to experimentally determine the properties of a unidirectional (UD) layer and calculate the full laminate properties based on this. Applying this to NCF materials is possible but requires the properties of the UD NCF layers. This however neglects the influence of fiber waviness and stitching yarn gaps due to the textile manufacturing process. This typically affects the compressive properties more severely than the tensile properties and can be accounted for by knock down factors [Asp04, Edg05].

Due to the triaxial nature of the applied NCF material this approach could not be applied directly. Instead the elastic properties of a UD layer were initially determined based on analytical rules of mixture and later calibrated based on bending tests of the full NCF composite lamina. The strength properties were taken from the literature and then calibrated using three point bending (3PB) test of the lamina. The interlaminar behavior was determined based on literature values and then calibrated using interlaminar shear test (ILS).

5.2.1 Mechanical behavior of CFRP

Laminated CFRP is made of different layers of carbon fiber reinforced plastic. One distinguishes between the intralaminar and interlaminar material behavior. Both are required to describe the full mechanical response. The intralaminar behavior describes the in-plane response of the laminate. In case of UD fiber reinforced composites, this is dominated by the fiber itself and thus must be described depending on the type and direction of the loading. The interlaminar behavior describes the out-of-plane response of the laminate which is dominated by the resin rich interface in between the individual plies. Failure of this interface leads to delamination of the individual plies.

CFRP responds brittle when loaded in the direction of the fibers – also referred to as fiber direction. Stiffness, strength and elongation at failure depend mostly on the fiber itself as it carries the majority of the load. In tension failure takes place very abruptly once the strength of the carbon fiber is superseded. In compression local fiber buckling leads to the formation of kink bands, bands of broken fibers that finally yield failure as they grow across the laminate. This failure mode is a local stability failure influenced by strength and stiffness of fiber and matrix as well as fiber geometric properties such as waviness [Pin06]. The material behavior perpendicular to the fibers – also referred to as matrix direction – is also brittle but characterized by an interaction of fiber and matrix properties. Matrix materials typically have a greater elongation at break than carbon fibers and thus allow more plasticity. As stiff fibers are added to the matrix, they act as a notch reducing the effective matrix strength. Depending on the loading different kinds of matrix cracks may emerge. Shear loading of a UD ply typically leads to greater plastic deformation before ultimate failure. As the damage mechanics during shear failure are – when looked at in detail – very complex, the exact description of the full stress strain curve is very difficult and subject of research [Pin06a]. Figure 5.4 shows a polished cut image of a composite laminate with impact damage that highlights the discussed failure modes and how they may interact.

In engineering applications composites are typically subject to multiple loads. Different criteria have been developed to describe failure of FRP materials. This topic is until today subject of research. A World Wide Failure Exercise (WWFE) was initiated in the 1990's and setup as an open competition between researchers to discuss the quality of failure criteria for fiber reinforced polymers. Its first part (WWFE-I) investigated the intralaminar behavior and concluded in 2004 [Hin04]. The results of the WWFE-I led to improvements of failure models that have been used without changes since the 1950's and 1960's. Furthermore the modified theories are currently implemented into numerical software packages [Hin11]. However also major shortcomings were found one being failure under three dimensional (3D) stress states. This became subject of the WWFE-II, which investigated triaxial stress states and concluded in 2013 [Kad13]. Other shortcomings such as failure of damaged laminates and fatigue may be addressed in future parts of the WWFE [Hin07, Kad07].

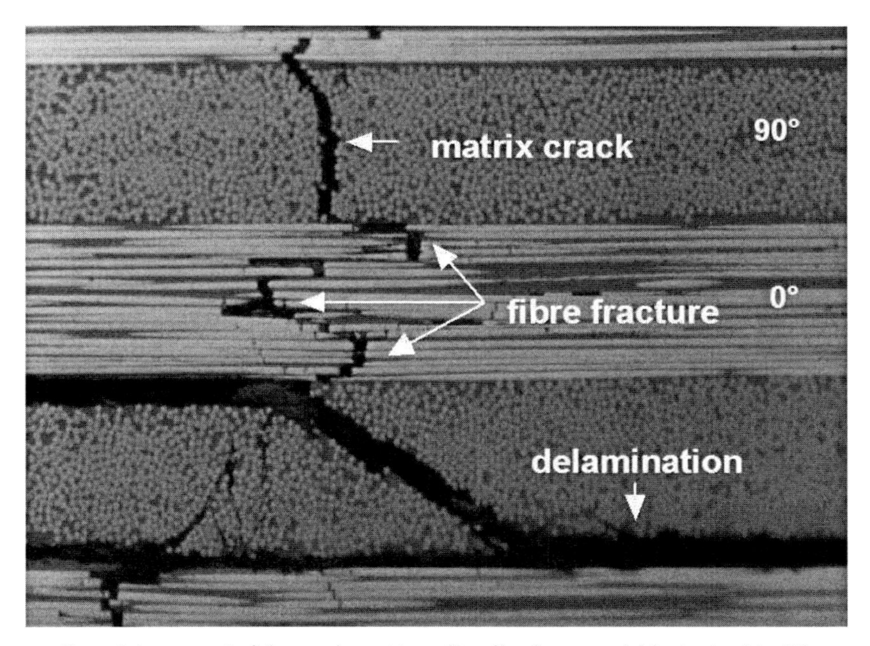

Figure 5.4 – composite failure modes matrix cracking, fiber fracture and delamination [Men98]

The extent of the WWFE is however much larger and it can thus not be discussed in detail in this work. Only a brief comprehension of relevant methods for the description of failure of composite materials will be provided in the context of their applicability to the numerical simulation of low velocity impact on CFRP. Generally failure criteria and material models for composites are grouped into different categories:

- max. strain failure criteria
- max. stress failure criteria
- global fracture failure criteria
- phenomenological based failure criteria
- progressive failure models
- continuum damage models

The categories will be explained and discussed using the example of a unidirectional composite as most failure criteria and models were developed specifically for this material type.

Maximum stress and strain failure criteria are the simplest way to describe failure of composites and are based on experimentally determined values. These criteria are reasonably well to describe failure in the fiber direction but lack accuracy once the material is subject to combined loadings of normal and shear stresses. Maximum strain criteria are often used for sizing structures subject to fatigue loads in order to limit matrix cracking.

Brittle fracture criteria describe material failure but do not distinguish between different modes. They can describe failure based on either stresses or strains. Most criteria use a two dimensional (2D) stress state within the lamina and describe a volume in the σ_{11}, σ_{22} and τ_{12} stress domain thus assuming the plane stress state. Stress states that are within the described volume do not cause failure while stress states outside of the volume lead to failure of the affected ply. Similarly a 3D stress state can be used that takes the effect of out-of-plane stresses into account. The first example of these models is Tsai-Hill [Hil48], which was initially developed for anisotropic metallic materials but then applied to composites. Other examples are Hoffmann [Hof67] and Tsai-Wu [Tsa72].

Phenomenological failure criteria describe particular damage modes of the composite material individually such as fiber failure in tension, in compression or matrix cracking. Safety of the complete structure can be determined by checking all relevant load cases against all possible failure modes. Hashin's [Has80] phenomenological failure criterion was one of the first of its kind and distinguishes between tensile and compressive failure modes in the fiber direction (fiber mode) and perpendicular to it (matrix mode). A piecewise smooth failure surface is created in the σ_{11}, σ_{22} and τ_{12} stress domain which encompasses a volume of safe design space. Research then concentrated on improving the individual failure surfaces that each describe a single failure mode leading to more physically based but also complex failure criteria such as Chang and Chang [Cha87].

More advanced criteria are Puck's interfiber fracture criterion [Puc96, Puc98] or Cuntze's [Cun97] failure mode concept with an application to thick wall laminates shown exemplary by Kroll and Hufenbach [Kro97]. Both rely on similar assumptions based on the Mohr-Coulomb fracture criterion for matrix failure due to compression shear loads. Puck uses a stress based description of the interfiber fracture plane while Cuntze applies an invariant formulation for the same. Recent research focuses on fiber compressive failure which is characterized by kink bands. These may originate due to locally misaligned fibers and lead to local fiber stability failure or matrix failure resulting in kink bands [Pin06].

A simple failure criterion is however insufficient for the simulation of short time dynamic processes that include material failure due to impact or crash. The mechanical response of FRP materials during damaging events requires the description of the complete failure process by e.g. brittle, plastic or fracture mechanics dominated behavior. In consequence two types of material models that account for failure have been developed.

Progressive failure models – also referred to as ply discount method – rely on phenomenological failure criteria which describe material failure of a single ply within a laminate. Typically the material is described as elastic until failure which then leads to immediate loss of the relevant material stiffness. In case of matrix cracking e.g. the transverse stiffness and shear stiffness of the plies may be degraded but the properties parallel to the fiber remain intact until maximum strength in this direction is reached. These criteria may then be extended by a simple formulation of the material law that describes post failure response by e.g. an elasto-plastic material law until a defined failure strain is reached which then leads to element erosion and thus complete material failure. True progressive failure can only be reached on the laminate level as the remaining plies have to bear the additional load of the failed ply.

Continuum or continuous damage models differ from progressive failure models in the use of fracture mechanics based material laws for the mostly brittle but also elasto-plastic behavior of fiber reinforced composites. Damage is included in a nonlinear stiffness degradation function which typically requires the definition of damage variables for each failure mode that run from e.g. 0 in the undamaged state to 1 in the fully damaged state and degrade the applicable stiffness. Examples of continuum damage models have been published by e.g. Ladeveze and Le Dantec [Lad92], Barbero [Bar01], Dávila and Camanho [Dav03], Pinho et al. [Pin06a], Puck [Puc07] and Maimí, Camanho et al. [Mai07]. Fatigue damage due to cyclic loading is typically not included in continuum damage models.

Within this work both a progressive failure model and a continuum damage model were investigated and applied. The progressive failure model was developed by Chang and Chang [Cha87] and serves as a reference while the continuum damage model was developed by Maimí, Camanho et al. [Mai07]. The Chang-Chang model uses a set of phenomenological failure criteria and initially proposed a property degradation model using an exponential function coupled with a critical fracture area. This was however modified as Matzenmiller and Schweizerhof [Mat91] implemented the model into LS-DYNA as material 54 (MAT54) "Enhanced Composite Damage". Property degradation was essentially replaced by an elastic-plastic material response that with strain controlled element erosion. This incorporates the idea of energy dissipation without however a direct relation to fracture mechanics. The Chang-Chang model was initially developed to describe damage in CFRP from stress concentrations such as open holes, bolts or impact damage.

The Chang-Chang failure model distinguishes between four principal failure modes. These are fiber tensile and compressive failure as well as matrix tensile and compressive cracking. Fiber failure is described using normal stresses in the fiber direction and the in-plane shear stress based on an expression of Yamada and Sun [Yam78]. Matrix cracking – also referred to as matrix failure mode – is described by the normal stresses perpendicular to the fiber orientation and in-plane shear stresses. After implementation into LS-DYNA a few changes have been made leading to the following numerical implementation by Livermore Software Technology Corporation (LSTC) [Lst06, Lst13, Lst13a]. Fiber failure is reached if one of the failure criteria ϕ_N described in equations (5.11) and (5.12) becomes greater or equal 1. Here β_S is a shear coupling parameter. These failure criteria are

$$\phi_{\mathrm{fib,T}}^2 = \left(\frac{\sigma_{11}}{X_T}\right)^2 + \beta_s\left(\frac{\tau_{12}}{S_L}\right) \quad \text{and} \tag{5.11}$$

$$\phi_{\mathrm{fib,C}}^2 = \left(\frac{\sigma_{11}}{X_C}\right)^2 . \tag{5.12}$$

Matrix failure is described by

$$\phi_{\mathrm{mat,T}}^2 = \left(\frac{\sigma_{22}}{Y_T}\right)^2 + \left(\frac{\tau_{12}}{S_L}\right)^2 \quad \text{and} \tag{5.13}$$

$$\phi_{\mathrm{mat,C}}^2 = \left(\frac{\sigma_{22}}{2S_L}\right)^2 + \left[\left(\frac{Y_c}{2S_L}\right)^2 - 1\right]\left(\frac{\sigma_{22}}{Y_C}\right) + \left(\frac{\tau_{12}}{S_L}\right)^2 . \tag{5.14}$$

Here X_T and X_C are the tensile and compressive strength in the fiber direction of the UD ply while Y_T and Y_C are the same in the matrix direction and S_L the in-plane shear strength. Chang and Chang [Cha87] propose a property degradation model which is also implemented into MAT54. Matrix failure reduces the in-plane shear and transverse material properties E_{22}, G_{12} and the related Poisson ratio ν_{12} to zero while the properties in the fiber direction – most notably E_{11} – are unaffected [Lst13a]. After fiber failure is reached, plastic behavior continues until the user defined maximum strain is reached. These strains are defined separately for tension and compression and lead to elimination of the affected ply from the laminate stiffness matrix. Additionally maximum transverse and shear strains can be defined to avoid excessively large element distortions due to transverse and shear strains. The resulting elasto-plastic material law is shown exemplary in Figure 5.5.

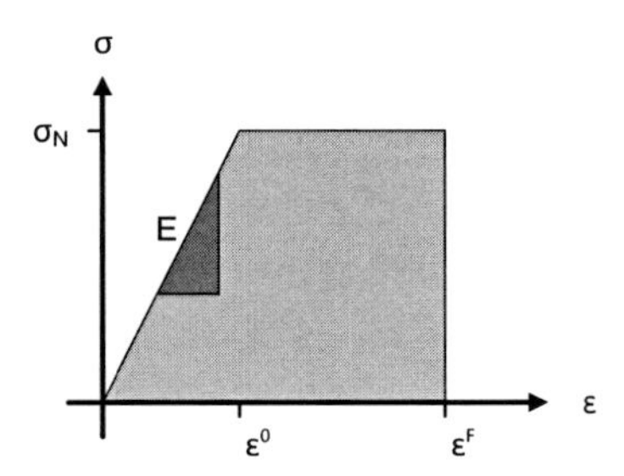

Figure 5.5 – MAT54: Elasto-plastic material behavior with failure

Here σ_N is the failure stress of the relevant failure mode N, ε^0 the failure initiation strain and ε^F the ultimate strain leading to element erosion. If all plies of a lamina are eliminated, the complete element will be eroded. Application of this stepwise material degradation leads to localized but stepwise softening of the material prior to complete laminate failure.

The second material model investigated was developed by Maimí, Camanho et al. [Mai07, Mai07a] and implemented recently into LS-DYNA as material 262 (MAT262) "Laminated Fracture Daimler Camanho" [Lst13a]. It distinguishes in principal between the same failure modes as MAT54, but applies more evolved failure criteria ϕ_N. These describe failure surfaces which together limit the elastic material behavior. Further they are used as damage activation functions, which lead to stiffness and strength degradation. Damage evolution is fracture mechanics controlled based on fracture toughness and interaction of the different failure modes. MAT262 distinguishes between tensile and compressive fiber failure using equations (5.15) and (5.16). For simplicity the same notation based on the true stresses σ and τ is used as previously for MAT54 while MAT262 actually applies the effective stresses $\tilde{\sigma}$ and $\tilde{\tau}$. Fiber failure in tension and compression is described by

$$\phi_{\text{fib,T}} = \phi_{1+} = \frac{\sigma_{11}-\upsilon_{12}\sigma_{22}}{X_T} = 1 \quad \text{and} \tag{5.15}$$

$$\phi_{\text{fib,C}} = \phi_{1-} = \frac{\langle|\tau_{12}^m|+\mu_L\sigma_{22}^m\rangle}{S_L} = 1 \quad \text{with} \tag{5.16}$$

$$\mu_L = -\frac{S_L\cos(2\alpha_0)}{Y_C\cos^2(\alpha_0)} \ ,$$
$$\sigma_{22}^m = \sigma_{11}\sin^2(\varphi^c) + \sigma_{22}\cos^2(\varphi^c) - 2|\tau_{12}|\sin(\varphi^c)\cos(\varphi^c) \ ,$$
$$\tau_{12}^m = (\sigma_{22}-\sigma_{11})\sin(\varphi^c)\cos(\varphi^c) + |\tau_{12}|(\cos^2(\varphi^c)+\sin^2(\varphi^c)) \ ,$$
$$\varphi^c = \tan^{-1}\left(\frac{1-\sqrt{1-4\left(\frac{S_L}{X_C}+\mu_L\right)\frac{S_L}{X_C}}}{2\left(\frac{S_L}{X_C}+\mu_L\right)}\right) \quad \text{and} \quad \alpha_0 = 53 \pm 2° \ .$$

Tensile fiber failure is thus based on fiber strength while compressive fiber failure is controlled by the matrix strength in the fracture plane. An initial fiber misalignment of φ^c and the fracture plane friction coefficient μ_L are based on strength properties. The upper script m denotes transformation of the global stresses to the matrix fracture plane which supports the fibers prior to kinking. Here α_0 describes the inclination of the compressive matrix fracture plane which was found to be equal to $53 \pm 2°$ for most engineering FRP materials [Puc98, Puc02]. Thus fiber compressive failure is essentially based on failure of the supporting matrix. It neglects however a local stability failure mode due to insufficient elastic support of the matrix as described by Pinho et al. [Pin06]. Matrix failure in MAT262 occurs if one of the two criteria in equations (5.17) or (5.18) are fulfilled:

$$\phi_{\text{mat,T}} = \phi_{2+} = \begin{cases} \sqrt{(1-g)\frac{\sigma_{22}}{Y_T} + g\left(\frac{\sigma_{22}}{Y_T}\right)^2 + \left(\frac{\tau_{12}}{S_L}\right)^2} = 1 & \text{if } \sigma_{22} \geq 0 \\ \frac{\langle|\tau_{12}|+\mu_L\sigma_{22}\rangle}{S_L} = 1 & \text{if } \sigma_{22} < 0 \end{cases} \tag{5.17}$$

$$\text{where} \qquad g = \frac{G_{Ic}}{G_{IIc}} \ .$$

Here g is the fracture toughness ratio of the matrix material. Matrix cracking perpendicular to the laminate plane is thus controlled by a combination of normal and shear loads. Compressive loads perpendicular to the fibers lead to matrix fracture in the fracture plane α_0. This fracture is described by the Mour-Coulomb fracture criterion as first explored by Puck et al. [Puc98]. MAT262 uses the following expression

$$\phi_{\mathrm{mat,C}} = \phi_{2-} = \sqrt{\left(\frac{\tau_t}{S_t}\right)^2 + \left(\frac{\tau_L}{S_L}\right)^2} = 1 \quad \text{if } \sigma_{22} < 0 \quad \text{with} \tag{5.18}$$

$$\tau_t = \langle -\sigma_{22}\cos(\alpha_0)\,[\sin(\alpha_0) - \mu_t\cos(\alpha_0)\cos(\theta)] \rangle \ ,$$
$$\tau_L = \langle \cos(\alpha_0)\,[|\tau_{12}| + \mu_L\sigma_{22}\cos(\alpha_0)\sin(\theta)] \rangle \ ,$$
$$S_t = Y_C\cos(\alpha)\left[\sin(\alpha) + \frac{\cos(\alpha_0)}{\tan(2\alpha_0)}\right] \ , \qquad \mu_t = -\frac{1}{\tan^2(2\alpha_0)} \ ,$$
$$\theta = \tan^{-1}\left(\frac{-|\tau_{12}|}{\sigma_{22}\sin(\alpha_0)}\right) \qquad \text{and} \qquad \alpha_0 = 53 \pm 2° \ .$$

Here τ_L and τ_t are the longitudinal and transverse shear stresses in the fracture plane while μ_L and μ_t are similarly the friction coefficients of the same. The angle θ now describes the orientation of the maximum shear stress in the fracture plane.

Degradation of the material properties is controlled by the fracture toughness of the CFRP corresponding to the individual failure mode. Maimí, Camanho et al. [Mai07a] propose linear-exponential damage evolution for longitudinal (fiber) failure and exponential damage evolution for transverse (matrix) failure. The implementation into LS-DYNA applies however a simpler bilinear damage evolution for fiber failure and linear damage evolution for matrix failure. Additionally a one dimensional elasto-plastic formulation with strain hardening is applied to account for the characteristic in-plane shear behavior of fiber reinforced plastics. The resulting material laws are shown in Figure 5.6. Here G_{nc} is the fracture toughness of the corresponding material failure mode and l_{el} the characteristic element length. For more details refer to appendix B.4 and LS-DYNA [Lst13a].

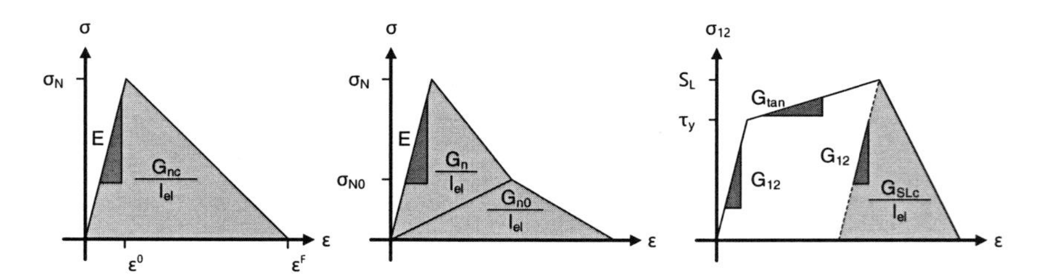

Figure 5.6 – material laws of MAT262: Linear damage evolution (left), bi-linear damage evolution (center) and elasto-plastic shear behavior coupled with linear damage evolution (right)

The interlaminar behavior of laminated CFRP describes the characteristics of the interface between two laminated plies. This behavior is dominated by the matrix, but also influenced by the characteristics of the individual plies. If the thin layer of matrix material between two plies fails, a crack will emerge which is referred to as delamination. A delamination can become critical as it is difficult to detect during visual inspection but leads to severe losses of the laminate bending and shear strength [Cam03].

Fracture mechanics principals can be applied to describe damage in the interface between two layers of material similar to those previously applied by the continuous damage model

MAT262 for the intralaminar behavior. In principal three crack opening modes are distinguished: Normal or tensile opening mode I, sliding mode II and tearing mode III – see Figure 5.7. Typically only modes I and II are of concern for laminated composites. In linear elastic fracture mechanics (LEFM) the stress field at the crack tip is characterized by the stress intensity factor K_n, with n being the applied opening mode. The fracture toughness that initiates critical crack growth is a material property [Gra03]. Depending on the fracture mode the corresponding material value is dubbed K_{nc} with c denoting critical.

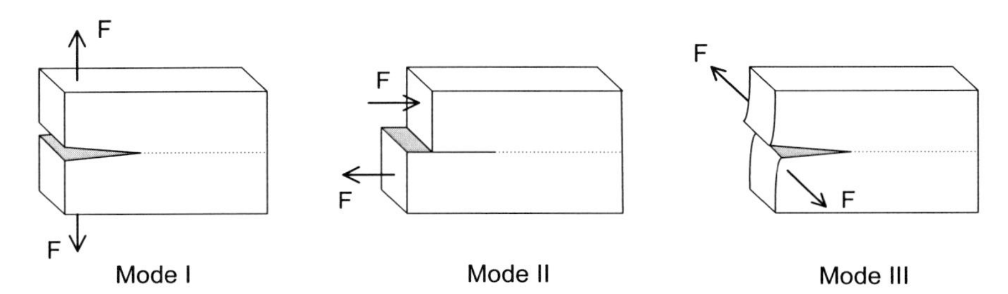

Figure 5.7 – crack opening modes: Normal (tensile) opening mode I, sliding mode II and tearing mode III

Similar to the stress intensity factor, the energy release rate G_{nc} can be used to describe crack growth. In LEFM – thus neglecting other dissipative effects such as material plasticity or fiber bridging – these two approaches are interchangeable. The energy release rate is based on the idea that the amount of energy required for crack growth is equal to the energy required to separate atomic bonds and form new surfaces [Roe08]. Using the energy release rate during crack growth in a particular fracture mode such as e.g. G_{Ic}, leads to the same result as the critical stress intensity factor and critical energy release rates can be transformed into each other by

$$G_{Ic} = \frac{K_{Ic}^{2}}{E'} \text{ with } E' = \begin{cases} E & : \text{for plane stress} \\ E/(1-\nu^2) & : \text{for plane strain} \end{cases} . \tag{5.19}$$

Energy release rates can be determined using the J-integral around the crack tip, which describes the energy released during crack growth. As the applied matrix materials are not purely brittle but also have some degree of plasticity, the application of LEFM is somewhat questionable. In consequence application of the energy release rate G has become more favorable, as it summarizes the energy required for a crack to form within a single value including dissipative effects not covered by LEFM such as e.g. plasticity and fiber bridging.

Current practice is to characterize a particular interface within the laminated CFRP by determining the critical energy release rates of modes I and II and their interaction, also referred to as mode mixity [Cam03]. For this purpose Double Cantilever Beam (DCB), End Notched Flexure (ENF) and Mixed Mode Bending (MMB) tests of the particular material are performed in order to determine the critical energy release rates G_{Ic} and G_{IIc} and their interaction as shown in Figure 5.8.

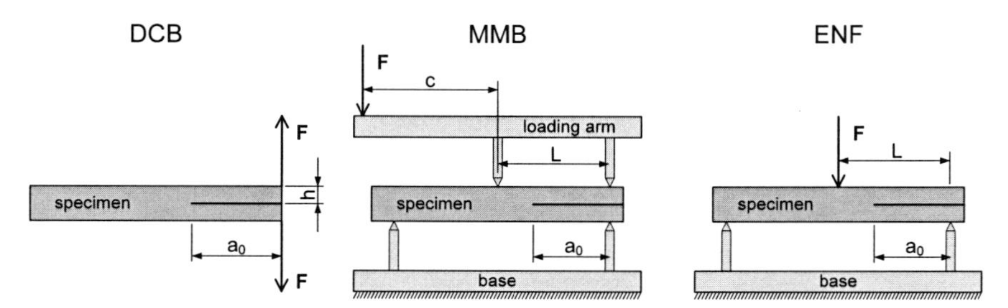

Figure 5.8 – DCB, MMB and ENF tests used for determination of the interlaminar behavior of CFRP

Numerical simulation of delamination of laminated composite materials is typically performed using either the Virtual Crack Closure Technique (VCCT) [Kru04] or cohesive elements [All95, Tur07]. The VCCT has been employed to describing the state of damage in structures made of composites [Kru01] or metal [Blo07]. It is however less effective in describing damage initiation and growth as it requires remeshing and accurate information from nodes both in front of and behind the crack tip.

The use of cohesive elements can overcome most of the difficulties related to the VCCT but at the cost of numerical difficulties including the proper definition of the stiffness of the cohesive layer, refined meshs across the affected areas and convergence problems related to softening constitutive models [Tur07]. Nevertheless the use of cohesive elements is by today implemented into most commercial finite element codes including LS-DYNA. In the view of the author cohesive elements are best suited for application to impact problems as they describe onset and propagation of delamination without remeshing.

Cohesive elements rely on the cohesive zone model which has already been applied to various fracture mechanics problems such as steel or concrete [Dug60, Hil76]. It assumes that damage develops in a cohesive zone ahead of the crack tip which thus loses part of its load carrying ability. The number of cohesive damage models applied to composite materials is large. The probably most common and relatively simple model is a linear triangular constitutive equation with mixed mode damage coupling as used e.g. by Camanho and Dávila [Cam03] and shown in Figure 5.9. Implementation into LS-DYNA has been performed by Gerlach et al. [Ger05]. Within LS-DYNA this model is available as material MAT138 "Cohesive Mixed Mode" in combination with cohesive elements or directly embedded within the tie-break contact algorithm.

The linear cohesive zone constitutive equation relates surface tractions τ to displacements Δ at an interface to the surrounding bulk material. The cohesive zone has the stiffness K_{cz} until the maximum traction (interfacial strength) τ^0 is reached. After the interfacial strength is reached, damage initiates and traction decreases linearly until maximum displacement Δ^F is reached. The energy dissipated during this is equal to the fracture toughness G_{nc}. This model is applied separately for opening modes I and II. Mixed mode interface problems can be described by the coupling parameter β_{cz} as shown in Figure 5.9.

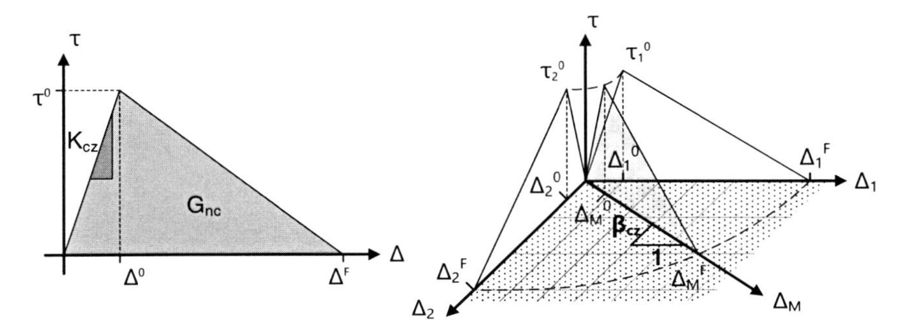

Figure 5.9 – constitutive equation (left) and coupling of damage modes (right)

One problem often encountered when using cohesive zone elements is the correct definition of the cohesive zone stiffness K_{cz}. This parameter, when chosen too high, can lead to spurious oscillations and numerical instabilities while it adds excessive compliance when chosen too low. For proper modeling of the cohesive zone typically at least three elements with a mesh size in the order of 1.0 mm or less are necessary to avoid mesh dependence of the interface response. Turon et al. [Tur07] describe an engineering solution to overcome mesh size effects when modeling delamination growth. At first the length of the cohesive zone l_{cz} is described by

$$l_{cz} = C_{cz} E_{cz} \frac{G_{nc}}{(\tau_n^0)^2} \quad \text{with} \quad C_{cz} = \frac{9\pi}{32} = 0.88 \,. \tag{5.20}$$

The parameter C_{cz} differs depending on the model of the cohesive zone length. Here a value of Falk et al. [Fal01] is used. Based on this formulation Turon et al. [Tur07] propose to artificially lower the interfacial strength τ^0 to a new value of τ^{0*} but keep the energy release rate G_{nc} constant which effectively increases the maximum displacement Δ^F by

$$\tau^{0*} = \sqrt{\frac{9\,\pi\,E_{cz}\,G_{nc}}{32\,n_{el}^0\,l_{el}}} \,. \tag{5.21}$$

Here n_{el}^0 is the number of elements used to describe the cohesive zone and l_{el} is the element length in the direction of crack growth. Using this approach has the effect of increasing the cohesive zone size and thus allowing the use of coarser meshes up to 3-4 mm size. This however poses problems as the artificially lowered interface strength τ^{0*} leads to premature crack initiation.

As there was only limited information available on the interfacial strength τ^0 for both opening modes n, equation (5.21) was inverted by putting in different values for n_{el}^0 and l_{el} in order to receive the correct surface strength required for describing crack growth. These values were then put into the FEM model of the ILS test and compared with the actual test data until τ_n^0 could be determined. Also the effect of varying element size l_{el} and number of elements n_{el}^{0} could be investigated for the performed material tests. Turon et al. [Tur07] also propose a mechanics based approach to calculate the cohesive zone stiffness K_{cz} with h_{ply} being the thickness of the neighboring sublaminates or plies

$$K_{cz} = \frac{\alpha_{cz} E_{33}}{h_{ply}} \quad . \tag{5.22}$$

This approach is based on the idea, that cohesive zone elements while being placed between two neighboring solid or thick shell elements have no thickness of their own. Selecting the parameter α_{cz} sufficiently large – Turon et al. [Tur07] propose a value of 50 – ensures, that the cohesive zone element does not add too much compliance to the model while remaining low enough to avoid numerical instabilities. As the simulation in this work is based on thin shell elements without compliance in the thickness direction, the proposed value of 50 turned out significantly too large. Instead the cohesive elements had effectively the thickness of a single ply bridging the gap between two plies in the laminate. Thus a value of $\alpha_{cz} = 1$ was used comparable to the approach of Airoldi et al. [Air13].

5.2.2 Material properties and strain rate sensitivity

The initial elastic material properties of the UD layers of the NCF composite were determined using rules of mixture implemented into the software Compositor [Ikv04]. These rules are based on straight fibers and do not take into account the effect of fiber waviness as it is found within NCFs. Typically the effect of fiber waviness can be provided for by knock down factors on the stiffness in the fiber reinforcement [Edg05]. As there was no information on the fiber waviness of the applied NCFs available, a knock down factor $\eta = 0.96$ was selected as proposed by Compositor [Ikv04]. The mechanical properties of the Toho Tenax HTS carbon fiber and the epoxy resin Hexcel RTM 6 were taken from the manufacturers data sheets [Hex12, Ten12] and are additionally summarized in appendix B. The resulting elastic properties are based on a nominal fiber volume content (FVC) of 60% typical for aerospace quality and given in Table 5.1.

Multiple experiments are required to determine five basic strength properties of a UD layer. These are the in-plane tensile and compressive failure stresses both parallel (X_T, X_C) and perpendicular (Y_T, Y_C) to the fiber direction and the in-plane shear strength (S_L). There is however only limited information available on the material combination used. Compositor provides detailed information on the properties of a combination of Toray standard modulus carbon fibers with the epoxy infusion resin LY556 [Ikv04]. Schubert [Sch09] describes the strength properties of a composite with an epoxy resin and Tenax HTS40 fibers based on a quasi-prepreg process eliminating textile effects of the NCF for comparison of an infusion resin system with a prepreg material. Hartung [Har09] provides test results of Tenax UTS fibers in a unidirectional NCF with Hexcel RTM-6 resin while Puck [Puc02] provides reference values for CFRP. Pinho et al. [Pin06a, Pin06b] perform a fracture mechanics based characterization of the prepreg T300/913 but also provide the bulk material properties. Table 5.2 summarizes the discussed material properties.

Table 5.1 – elastic properties of the NCF UD ply determined by rules of mixture

FVC [%]	ρ [g/cm^3]	E_{11} [MPa]	E_{22} [MPa]	G_{12} [MPa]	G_{23} [MPa]	ν_{12} [-]	ν_{23} [-]	$\alpha_{T,11}$ [K^{-1}]	$\alpha_{T,22}$ [K^{-1}]
60	1.52	139000	9600	4900	3950	0.278	0.22	0.4×10^{-6}	35×10^{-6}

Table 5.2 – strength properties of different CFRP UD materials

fiber	resin	product type	X_T [MPa]	X_C [MPa]	Y_T [MPa]	Y_C [MPa]	S_L [MPa]	source
Toray T300	LY556	textile	1500	900	27	200	80	[Ikv04]
Toray T700	LY556	textile	2125	1570	45	200	70	[Ikv04]
Tenax HTS40	epoxy	quasi-prepreg	2478	1520	79	-	71	[Sch09]
Tenax UTS	RTM-6	NCF	2208	1095	42	189	72	[Har09]
N.N.	epoxy	N.N.	-	-	50	230	100	[Puc02]
Toray T300	epoxy 913	prepreg	2005	1650	68	198	150	[Pin06a]

The strength properties from Schubert [Sch09] were used as a baseline for the fiber dominated specimens. As the tested specimens contained a higher FVC of 62 – 63%, a knock-down factor of 15% is applied to the fiber dominated properties X_T and X_C to account for the lower FVC and fiber waviness as a result of the NCF textile process. Maximum failure strain for tensile and compressive loads in the fiber direction (ε_T and ε_C) was chosen with a value of 1.8% identical to the pure Tenax HTS carbon fiber [Ten12].

The resin dominated properties Y_T and S_L are selected to 70 MPa and 100 MPa respectively. It is noted that all sources but Schubert provide lower values for Y_T compared to S_L which is thus accounted for. The value for Y_C is selected to 200 MPa. All values provided are an approximation based on the bulk material but can be used as a starting point for calibration of the material properties. Bulk material properties typically underestimate the in-situ material strength of the resin dominated material properties of a laminate as they do not account for stacking sequence and ply thickness effects [Cam06].

The interlaminar properties G_{Ic} and G_{IIc} were finally obtained from de Verdiere et al. [Ver12] based on interlaminar tests of a NCF material of the same combination of fiber and resin. Here a fracture toughness of 426 J/m^2 is reported for mode I and 1500 – 2000 J/m^2 for mode II. All preliminary strength properties are summarized in Table 5.3.

Table 5.3 – initial material strength properties of the NCF UD ply

X_T [MPa]	X_C [MPa]	Y_T [MPa]	Y_C [MPa]	S_L [MPa]	G_{Ic} [J/m^2]	G_{IIc} [J/m^2]	ε_T [%]	ε_C [%]
2100	1300	70	200	100	426	1500	1.8	1.8

Materials subject to short time dynamic loads with subsequent high strain rates may exhibit dynamic stiffening and strengthening but also embrittlement. This type of material characteristic is commonly referred to as strain rate sensitivity or simply rate sensitivity. In case of CFRP rate sensitivity is more complex due to the heterogeneous character of the composite material. Carbon fibers as base material are generally found to be not rate sensitive while in contrast the polymer matrix as the second base material is reported rate sensitive. In consequence rate sensitivity has to be discussed separately for different loads and subsequent failure modes of laminated CFRP. Measuring rate sensitive material properties accurately can be very challenging. For low and medium strain rates up to 0.1 s^{-1} typically servo hydraulic testing machines are used while tests at higher strain rates require more complex setups such as e.g. the split hopkinson pressure bar [Har83].

Strain rate sensitivity of CFRP is widely discussed in the literature as the review article of Sierakowski [Sie97] demonstrates. It is generally accepted that the tensile properties of UD CFRP loaded in the fiber direction – also referred to as longitudinal properties – are not rate sensitive, while tensile properties with a material orientation other than this are rate sensitive [Har83, Tan07, Zho07a]. Compressive and shear properties in the fiber and matrix directions are also rate sensitive [Hsi98, Hos01, Bin05, Koe11, Wie08, Koe11a]. In summary all material properties except the longitudinal tensile properties are reported as rate sensitive. The type and extend of the rate sensitivity varies however significantly depending on the investigated material and reference.

Quantifying rate sensitivity depends significantly on the selected material combination as the polymeric matrix is rate sensitive. Only limited influence of fiber waviness is reported that could be related to manufacturing methods [Hsi98]. As there are – to the knowledge of the author – in the literature no results available for the combination of Tenax HTS carbon fibers and RTM-6 epoxy resin, an engineering judgment was performed using test results from similar but not identical materials.

Gerlach et al. [Ger08] investigated the rate sensitivity of neat Hexcel RTM-6 resin. It is reported that viscoelastic behavior of the resin and thus rate sensitivity becomes noticeable for strain rates larger than 1 s^{-1}. For strain rates in excess of 1000 s^{-1} resin stiffness triples while strength nearly doubles compared to the quasi-static values. The material thus becomes more brittle and maximum failure strain drops noticeably.

Looking at CFRP, Harding and Welsh [Har83] as well as Zhou et al. [Zho07a] investigated the longitudinal tensile properties and thus report no noticeable rate sensitivity. Taniguchi et al. [Tan07] report the same for the longitudinal tensile properties but describe significant rate effects for the tensile properties perpendicular to the fiber – also referred to as transversal properties – and for the shear properties. Both are subject to increasing stiffness. In case of the transversal tensile properties the ultimate strength increases only moderately by up to 25% for rates larger than 100 s^{-1} while in case of the shear strength an increase up to a factor of two is reported for the same strain rates.

Reported rate effects of the compressive material properties vary stronger in the literature. Hsiao and Daniel [Hsi98] report for the longitudinal properties only little influence on the stiffness but an increase of up to 75% of the compressive strength for strain rates larger than 100 s^{-1}. The transversal and shear compressive stiffness is reported to increase by about 25% while the strength increases 75-100% for rates larger than 100 s^{-1}. Maximum strain is reported as nearly insensitive to strain rates and is suggested as a failure criterion. Hosur et al. [Hos01] report similar trends and smaller increases in strength and stiffnesses but describe a reversal of the trend for strain rates larger than 200 s^{-1}. Bing and Sun [Bin05], Wiegand [Wie08] and Koerber et al. [Koe11, Koe11a] report similar trends but with somewhat varying numbers. Wiegand [Wie08] provides a comprehensive comparison of the described and additional literature sources in addition to test results of his own.

Based on the above described literature sources an engineering approach is applied for scaling the five CFRP quasi-static base strengths X_T, X_C, Y_T, Y_C and S_L for strain rates up to 10^4 s^{-1} in order to cover the most relevant rate effects. The quasi-static values are referred to as reference value and all scale factors are shown in Figure 5.10. A similar approach could be applied to describe rate sensitive stiffnesses behavior of the CFRP face sheet but was omitted due to a lack of implementation into LS-DYNA. The full rate sensitive material properties are summarized in appendix B.7.

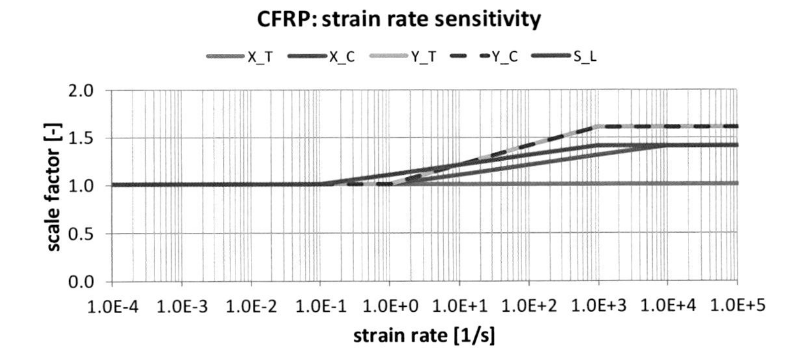

Figure 5.10 – strain rate dependent scaling of CFRP strength properties

5.2.3 Experiments

Separate experiments have been performed with the face sheet material for validation of applied material models and calibration of material properties. For this purpose skin material of surplus sandwich specimens was stripped from the foam core and investigated in three-point bending (3PB) and interlaminar shear (ILS) tests. The striped monolithic CFRP was taken from the top face sheet of the sandwich as this side is hit by the impactor.

Using the striped skin material instead of separately manufactured monolithic CFRP has advantages and disadvantages. The main advantage is that effects from the manufacturing process such as the flexibility of the foam core and heat distribution during the manufacturing process are identical to the sandwich structure. The disadvantage is that the striped CFRP specimens have a thin film of residual resin and foam on their lower side which dates back to the previous foam core interface. This residue was minimized by grinding but could not be removed completely.

The influence of the interface residue on the 3PB test is difficult to quantify. Qualitatively the residue leads to a greater specimen thickness and thus increased bending stiffness. The stiffness of the pure resin is however one dimension smaller than that of the reinforced plies which reduces the effect of the additional thickness on the bending stiffness. Additionally the interface residue introduces a limited laminate unbalance.

The 3PB tests were performed according to DIN EN ISO 14125 with $R_1 = 5$ mm, $R_2 = 2$ mm, $L = 80$ mm and a specimen width $b = 15$ mm. Figure 5.11 displays the principal test setup. A preliminary test force of 25 N is applied as reference position prior to the test force. Tests were performed in the 0° and 90° orientation of the laminate using a face sheet with $h = 1.5$ mm nominal thickness and a $[((45°/0°/135°)_s)_2]$ stacking sequence. This deviates from the test standard as it requires a thickness of $h = 2$ mm. The interface residue led however to a measured average specimen thickness of 2.08 mm in case of the 0° specimens and 2.19 mm in case of the 90° specimens [Sal11].

The lower specimen thickness decreases the specimens bending thickness. As the test results are however only used as experimental reference for a simulation model that applies the true specimen thickness, the deviation from the test standard is not considered as critical. The tested 0° specimens failed either by compressive or tensile fracture in the outer layers while the 90° specimens failed only by tensile fracture. No interlaminar shear failure was recognized. Figure 5.12 shows the load vs. displacement curves of all test specimens.

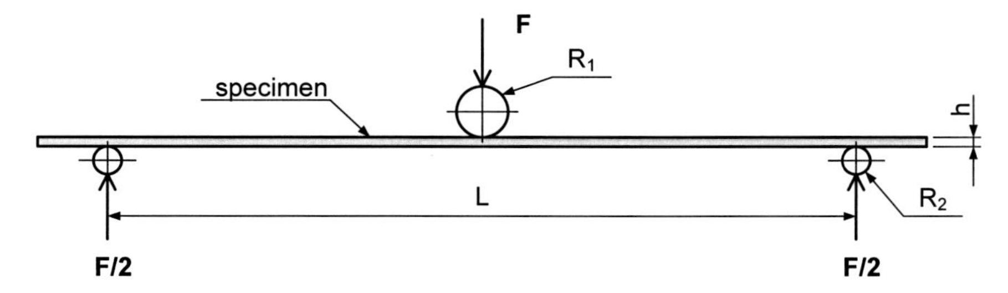

Figure 5.11 – three point bending (3PB) test setup of for monolithic CFRP

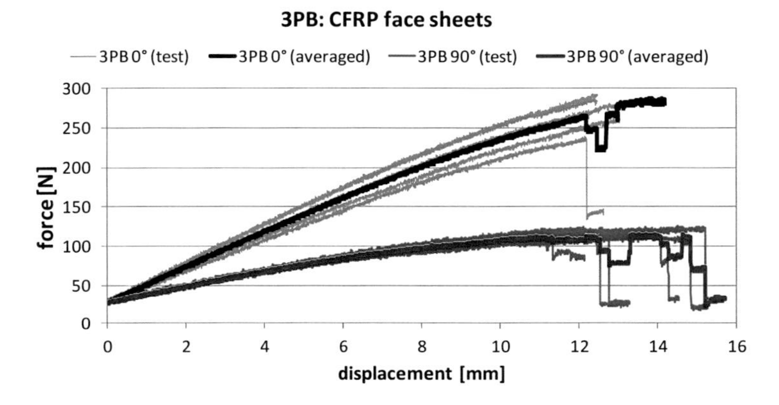

Figure 5.12 – results of 3PB tests of the CFRP face sheet in 0°- and 90° orientation; averaged values are shown in black (0°) and grey (90°) color

The 3PB test results are in agreement with expectations. The main load carrying part of the laminate are the layers with a fiber orientation parallel to the in-plane loads. If one of these highly loaded layers fails, the whole specimen will likely fail completely as the other layers are not capable of bearing the additional load. As the compressive strength of non crimp fabrics can be – depending on their individual stitching pattern and related fiber waviness – as low as half of their tensile strength [Asp04], primarily compressive failure is expected. It is however difficult to judge this as during specimen failure secondary damages on the tensile loaded bending side occur, too. Also an influence of the interface residue on the lower (tension) side of the specimen is possible, but was not observed.

The ILS tests were performed according to the standard DIN EN 2563. Figure 5.13 displays the principal test setup with $R_1 = 3$ mm, $L = 10$ mm and a specimen width $b = 10$ mm. A preliminary test force of 25 N is applied as reference position prior to the test force. Tests were performed both in the 0° and 90° direction of the laminate using a face sheet with the nominal thickness $h = 1.5$ mm and a $[((45°/0°/135°)_s)_2]$ stacking sequence. This deviates from the test standard which requires a thickness of $h = 3$ mm. The measured average specimen thickness was 2.09 mm in case of the 0° specimens and 2.18 mm in case of the 90° specimens [Sal12] showing the effect of the interface residue similar to the 3PB test. Figure 5.14 shows the load vs. displacement plots of all test specimens and their averaged value.

General shortcomings of the ILS test are a nonlinear load distribution and interaction of matrix cracking and delamination. The thinner test specimens increase these as in-plane loads become greater than envisaged by the test standard. Failure occurred in all specimens due to delamination. No signs of fiber failure or matrix cracking were detected visually. The test results in Figure 5.14 show a noticeable nonlinearity at the beginning which is attributed to indentation of the residual foam core interface on the lower specimen side.

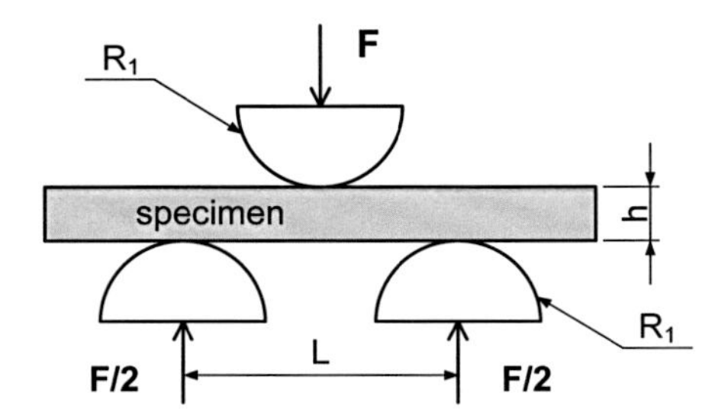

Figure 5.13 – interlaminar shear (ILS) test setup for monolithic CFRP

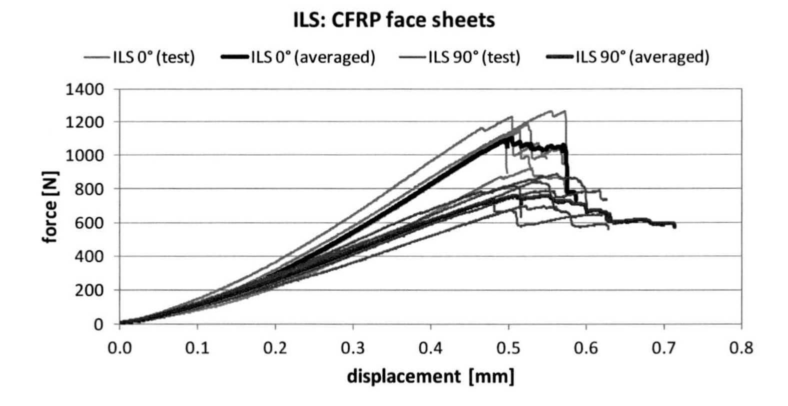

Figure 5.14 – results of ILS tests of the CFRP face sheet in 0°- and 90° orientation; averaged values are shown in black (0°) and grey (90°) color

5.2.4 Simulation with one element across the thickness

Simulation of the CFRP face sheet can be done at different levels of detail. A simple but common approach is to exclude interlaminar failure and model the face sheet as a stacked shell element. Alternatively a more sophisticated approach using one shell element per ply can be used. The shell elements then have to be connected to each other to create the laminate. This can be done by e.g. cohesive zone elements for modeling interlaminar failure. This work investigated and applied both approaches towards their suitability to describe the failure behavior using the 3PB experiments as a test case.

Figure 5.15 shows the numerical model of the 3PB test used for validation of the material stiffness. For this purpose the CFRP laminate was discretized in the simulation with one stacked shell element across its thickness. The loading is applied using rigid supports modeled as volume elements. The central indenter moves with a defined velocity while the two outer supports are fully constrained.

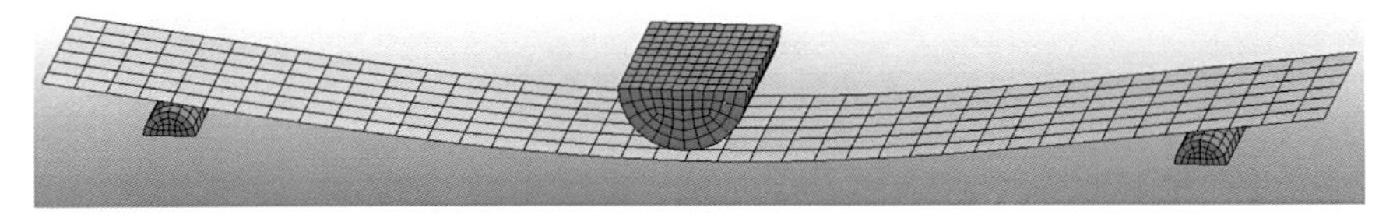

Figure 5.15 – simulation of the 3PB test with one shell element across the thickness

Two different shell element formulations were investigated. These are the default Belytschko-Tsay element formulation with only one integration point (LS-DYNA type 2) and a fully integrated shell element with 2 x 2 integration points (LS-DYNA type 16). The main advantage of the fully integrated element is that no zero energy deformation modes – typically referred to as hourglassing – can occur [Dyn12]. Additionally an improved accuracy is stated however without quantification, while the main disadvantage is a higher computational effort. The numerical cost penalty of element type 16 is quantified to a factor of 2.5 compared to element type 2 [Dyn12]. As hourglassing was observed in the sandwich model, element type 16 was chosen for all subsequent studies.

When using the explicit finite element method the global calculation time step is determined by the critical time step of the smallest element. This is typically on the order of magnitude of microseconds. Simulating a quasi-static test with the same displacement speed as applied in the experiment would thus lead to excessively huge numbers of calculation steps and related problems in terms of computational time as well as accuracy due to numerical rounding. Indenter displacement was thus raised from 0.08333 mm/s in the test to 1000 mm/s in the simulation. At this speed oscillations occur due to the load introduction or the occurrence of material damage as visible in Figure 5.16 and Figure 5.17. Oscillations in the simulation are therefore related to the high displacement velocity of the indenter and were not found in the experimental results.

Figure 5.16 and Figure 5.17 show force vs. displacement plots of the 3PB tests and simulation results. The simulation results are based on the material properties in Table 5.1 and Table 5.3 and compare different element sizes between 0.5 mm and 5 mm. The plots show that element size does not alter the bending stiffness in the simulation but has in case of the 0° orientation test a noticeable effect on specimen failure. This can be attributed to the well known mesh dependency of material damage parameters in finite element calculations [Ste09, Har09, Har12]. A failure strain of 1.8% is used in the fiber direction as stated in Table 5.3. At this point erosion of individual plies and their corresponding integration points will take place. The affected element will be erased once all plies of the laminate fail. Small elements contribute to element erosion as they describe stress concentrations such as e.g. the 3PB load introduction better. Thus small elements reach the failure strain earlier than larger elements which in contrast blunt stress concentrations. This can be noticed in Figure 5.16 as the maximum displacement reduces from about 12.5 mm for 5.0 mm elements to about 11 mm for 0.5 mm elements.

Figure 5.16 and Figure 5.17 also reveal that the simulated stiffness of the 3PB specimen is in the 0° orientation slightly greater than in the experiment. In the 90° orientation specimen failure was not simulated correctly as shown in Figure 5.17. Element erosion due to matrix failure is not applied thus limiting laminate failure to fiber failure only.

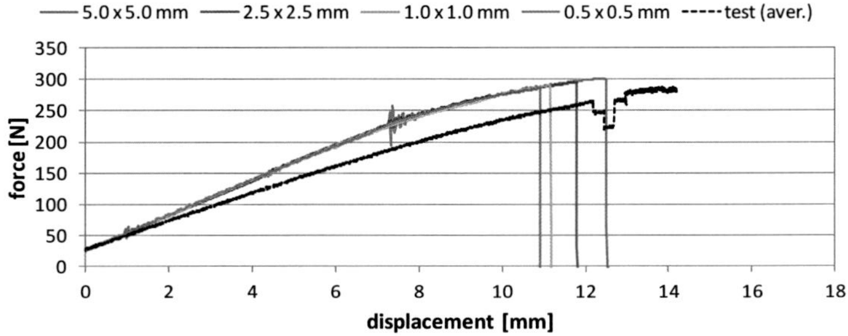

Figure 5.16 – comparison of 3PB test and simulation in laminate 0° orientation with different mesh sizes

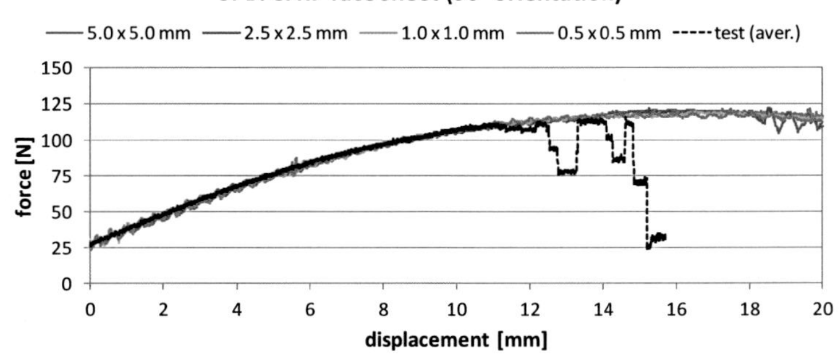

Figure 5.17 – comparison of 3PB test and simulation in laminate 90° orientation with different mesh sizes

Measurement of the FVC of the CFRP face sheet showed that the nominal value of 60% was not met [Met11]. Instead a mean value of 57.4% was measured for the upper face sheet. In consequence the nominal FVC was reduced to 58% which describes the tested laminate more closely and then used as input value to determine the elastic material properties using rules of mixture [Ikv04]. Additionally the knock down factor for fiber waviness, which affects only the UD stiffness in the 0° orientation, was changed from the initial guess of 0.96 [Ikv04] to 0.9. This aggress better with values for NCF composites reported by Asp et al. [Asp04]. Here it is stated that the tensile stiffness of an NCF drops between 3% and 10% compared to a prepreg material while the compressive strength degradation is reported to be on the order of 15% to 18%.

In order to stay consistent with the amount of dry fibers in the face sheets, the ply thickness had to be increased from nominally 0.125 mm to 0.133 mm in agreement with a FVC of 58%. The nominal thickness of the 3PB specimen thus increases similarly from 1.5 mm to 1.6 mm. Finally maximum failure strain in the fiber direction was raised from 1.8% to 2.2% in order to describe specimen failure correctly. It is noted that this failure strain is adjusted to an element size of 1 mm and thus not a true physical parameter. The revised material parameters are summarized in Table 5.4 and Table 5.5.

The performed adjustments reflect the known mesh dependency and the lack of in-plane test results for the UD plies of the applied triaxial NCF material. The revised stiffness properties now agree well with test results presented by Schubert [Sch09] and Hartung [Har09]. Figure 5.18 presents force vs. displacement of 3PB test and simulation with the initial material properties (sim_1) and the revised properties (sim_2) with a fixed element size of 1 mm. The bending stiffness increased due to the greater thickness of the 3PB specimen of now 1.6 mm and despite the reduced material stiffness.

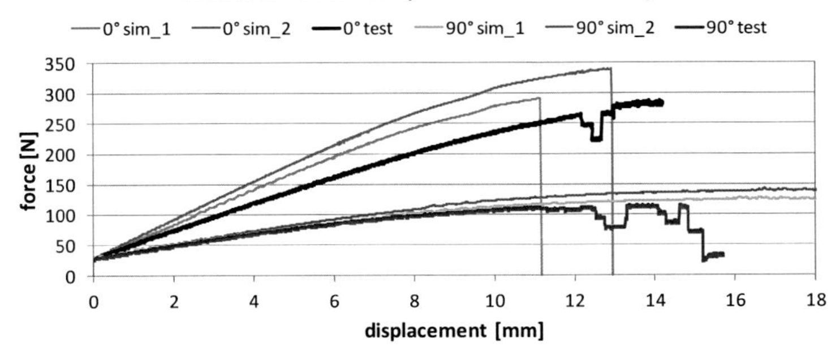

Figure 5.18 – comparison of 3PB test and simulation in laminate 0° and 90° orientations using the initial properties (sim_1) and the revised properties (sim_2) of the NCF UD ply

Table 5.4 – elastic properties of the NCF UD ply using one shell element across the thickness

FVC [%]	ρ [g/cm³]	h_{ply} [mm]	E_{11} [MPa]	E_{22} [MPa]	G_{12} [MPa]	G_{23} [MPa]	ν_{12} [-]	ν_{23} [-]
58	1.51	0.132	126000	9100	4600	3800	0.28	0.2

Table 5.5 – strength properties of the NCF UD ply using one shell element across the thickness

X_T [MPa]	X_C [MPa]	Y_T [MPa]	Y_C [MPa]	S_L [MPa]	ε_{XT} [%]	ε_{XC} [%]	ε_Y [%]	ε_{SL} [%]
2100	1300	70	200	100	2.2	2.2	20	20

5.2.5 Simulation with one element per ply

The second more detailed approach to model CFRP requires one shell element per ply that are connected to each other with cohesive elements. Here the shell elements describe intralaminar failure while the cohesive elements are responsible for debonding as interlaminar failure mode. This approach can now describe the correct kinematics of delamination as failed cohesive elements eliminate the mechanical connection of the affected plies. Validation of this approach against experimental data is more difficult as it includes intra- and interlaminar failure modes which typically affect each other by stress redistribution and local stress concentrations. One way to avoid this is to use the test results of the pure interlaminar tests DCB, MMB and ENF. As experimental results were limited to 3PB and ILS tests, instead a partially iterative process was adopted:

1. Simulation of 3PB test with one shell element across the thickness. Establishment of intralaminar material properties.
2. Determination of interlaminar material properties from literature results [Ver12] and use of recommendation from Turon et al. [Tur07].
3. Calibration of the interlaminar face sheet properties by modeling the ILS test with one shell element per ply connected by cohesive elements. Interlaminar failure is described by cohesive elements.
4. Calibration of intralaminar face sheet strength parameters by modeling 3PB tests with one shell element per ply. Interlaminar behavior is described using cohesive elements based on material properties from step 3.
5. Repeat steps 3 and 4 if necessary until stable intra- and interlaminar material properties are determined.

The first step has already been performed in the previous section thus the revised material properties of Table 5.4 and Table 5.5 are used as the baseline.

The next step involves the determination of material properties for the cohesive zone. Using equation (5.21) the length of the cohesive zone can be estimated which requires knowledge of the cohesive zone stiffness. In the literature different values are used as summarized by Turon et al. [Tur07], which are however often the result of numerical best practice. Generally the cohesive zone stiffness can be varied between two extremes without affecting the results noticeably. The upper limit is given by the stability of the numerical algorithm while the lower limit is instead given by physical reasoning as a very low stiffness introduces additional compliance to the model which is not found in the real structure.

Turon et al. [Tur07] propose to select the cohesive zone stiffness based on a fixed increase of fictitious compliance to the model, e.g. 2%. This increase of the compliance is determined based on the stiffness and thickness of the lamina surrounding the cohesive zone using equation (5.22). The proposed value of $\alpha_{cz} = 50$, which results in an increase of 2% to the total compliance, is however only true if thick shell or volume elements are used for the plies. These elements have already a stiffness and thus compliance assigned in the thickness direction. Thin shell elements as used in this model have however no compliance in the thickness direction. Thus the missing compliance of the thin shell elements is replaced by the compliance of the cohesive zone creating a true three dimensional laminate behavior as proposed by Airoldi et al. [Air13].

The interface strength of the HTS40/RTM-6 CFRP is however unknown. An educated guess is that the interface strength is less or equal to the tensile strength of the pure resin, which is given by the manufacturer as σ = 75 MPa [Hex12]. Nonlinear effects such as stress concentrations, fiber bridging and thermal stresses may however counteract this. Turon et al. [Tur07] provide for the prepreg material T300/977-2 CFRP an interface strength of τ_I^0 = 60 MPa. Camanho and Dàvila [Cam03] propose values of τ_I^0 = 80 MPa and τ_{II}^0 = 100 MPa for AS4/PEEK CFRP laminate. Hartung [Har09] performed tests with UTS RTM-6 CFRP specimens of the normal and shear interface strength on specifically designed through the thickness coupons and measured strength values of τ_I^0 = 36 MPa and τ_{II}^0 = 41 MPa. The apparent interlaminar shear strength τ_{II}^0 measured during the ILS tests was 41 MPa for the 0° oriented face sheet laminate and 27 MPa in the 90° orientation [Sal11].

Table 5.6 now shows the results for the cohesive zone length using equation (5.21) with a cohesive zone stiffness of K_N = 7.0*10^4 MPa/mm and K_T = 3.2*10^4 MPa/mm and different values of τ_n^{0*} between 30 MPa and 50 MPa. K_N and K_T were determined using equation (5.22) with α_{cz} = 1 and the material properties of Table 5.4. Depending on the chosen interface strength the length of cohesive zone varies between 1.28 mm and 5.74 mm. Focus will be set on opening mode II as this is the main loading during the impact and the only value that can be validated using the ILS tests.

Simulating the cohesive zone requires a fine mesh and results are typically dependent on the mesh size if the mesh becomes course. Turon et al. [Tur07] use typically five elements but states that other authors consider as few as two elements sufficient referring to Falk et al. [Fal01]. In order to weigh different options for the discretization of the cohesive zone, its properties were determined using equation (5.21) for different element edge length l_{el} and numbers of elements n_{el} that span across the cohesive zone. The values for τ_I^{0*} and τ_{II}^{0*} are required as input values for the numerical simulation. The element edge length is varied between 0.5 mm and 1.0 mm with all results summarized in Table 5.7.

Table 5.6 – length of the cohesive zone depending on the interface strength

	$l_{cz,I}$ [mm]; G_{Ic} = 426 J/m²	$l_{cz,II}$ [mm]; G_{IIc} = 1500 J/m²
$\tau_I^{0*} = \tau_{II}^{0*}$ = 50 MPa	1.37	2.21
$\tau_I^{0*} = \tau_{II}^{0*}$ = 40 MPa	2.14	3.45
$\tau_I^{0*} = \tau_{II}^{0*}$ = 30 MPa	3.81	6.13

Table 5.7 – cohesive zone strength depending on the element edge length and number of elements

	Mode I (τ_I^{0*})			Mode II (τ_{II}^{0*})		
Element edge length	$n_{el}= 5$	$n_{el}= 3$	$n_{el}= 2$	$n_{el}= 5$	$n_{el}= 3$	$n_{el}= 2$
l_{el} = 1.0 mm	26.17	33.79	41.38	33.21	42.87	52.50
l_{el} = 0.5 mm	37.01	47.79	58.53	46.96	60.63	74.25

Simulations of the ILS tests of the face sheet laminates both in the 0° and 90° orientation were performed using element lengths of 0.5 mm and 1.0 mm. The cohesive zone strength properties τ_I^{0*} and τ_{II}^{0*} were selected from Table 5.7 based on a discretization of five or three elements across the length of the cohesive zone as shown in Figure 5.19.

Numerical stability of the simulation was affected by the stiffness of the cohesive zone and the speed of the indenter. Particularly the model with an element edge length of 1.0 mm was subject to instabilities due to spurious oscillations which required a reduction of the time step to 70% of the critical value for a stable calculation. Simulation and test results are compared in Figure 5.20 and Figure 5.21.

From this comparison, it becomes clear that the simulation is generally capable of describing the interlaminar shear response but overestimates the interlaminar shear stiffness. At the beginning of the force vs. displacement curve the test results show a noticeable nonlinearity which also appears in the simulation. The nonlinear response of the simulation model does however fade out faster in comparison to the test results. This may be explained by the simplified simulation model which omitted foam core interface residue on the lower side of the face sheet specimens. All ILS test specimens were taken from striped sandwich face sheets and include limited residue of the foam core interface on the lower side of the laminate. Investigations of tested ILS specimens show that this interface residue deforms plastically during testing which explains the nonlinear response at the beginning of the test at least partially.

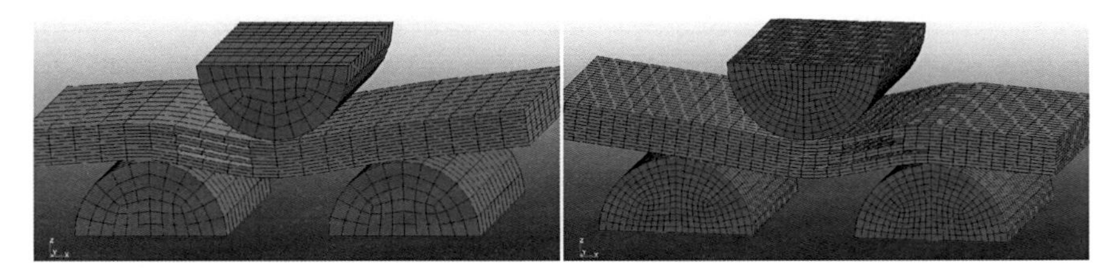

Figure 5.19 – simulation model of ILS test with element sizes of 1.0 mm and 0.5 mm with shell elements (grey) which represent the intralaminar properties; the shell elements are connected by cohesive elements (blue) that represent the interlaminar response and erode as failure occurs

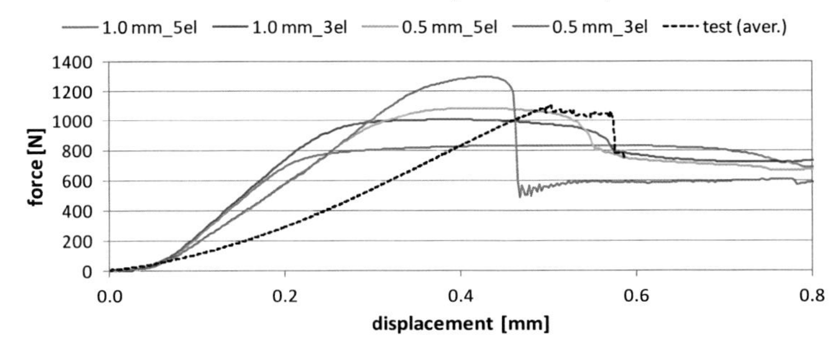

Figure 5.20 – simulation of ILS test in laminate 0° orientation with different element sizes (0.5 and 1 mm) and discretizations (3 and 5 elements) of the cohesive zone

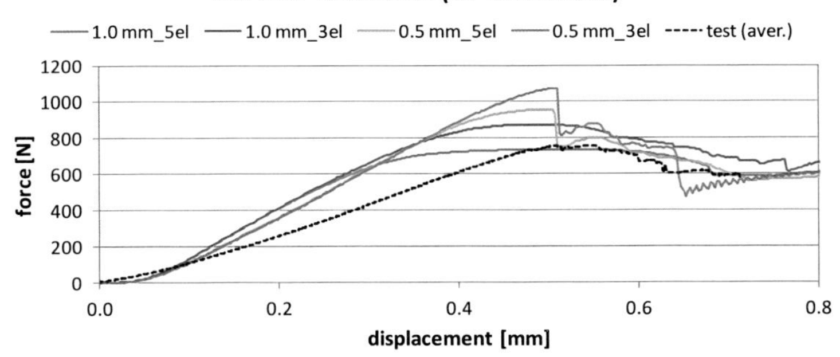

Figure 5.21 – simulation of ILS test in laminate 90° orientation with different element sizes (0.5 and 1 mm) and discretizations (3 and 5 elements) of the cohesive zone

Two observations were made when comparing the ILS test and simulation results of Figure 5.20 and Figure 5.21. First the stiffness overestimation is more severe in the 0° orientation than in 90°. The second is that for l_{el} = 1.0 mm and n_{el} = 3 the simulation overestimates the interlaminar strength in the 90° direction while it underestimates it in the 0° direction. This indicates an orthotropic behavior of the interlaminar properties not covered by the simulation. The simulation describes interlaminar stiffness solely by isotropic interlaminar cohesive zone elements. As the simulation utilizes thin shell elements, the out-of-plane laminate stiffness is modeled by isotropic cohesive elements and thus not capable to describe the orthotropic interlaminar shear stiffnesses G_{12} and G_{23} using instead an average value. This limitation may be overcome by thick shell elements which model out-of-plane laminate properties and reduce the function of cohesive zone elements to the interlaminar response. This approach however requires cohesive zone elements with very high stiffnesses to avoid too much fictitious compliance as discussed previously.

Failure of the ILS specimens is initiated by interlaminar shear failure during the simulation in agreement with the test results. The failure load depends as expected predominantly on the mode II cohesive zone strength τ_{II}^0. The most realistic failure load is reached using three elements across the length of the cohesive zone with an element size of 1.0 mm. This correlates with a mode II interlaminar strength of around $\tau_{II}^0 = 42$ MPa and agrees well with test results of Hartung [Har09] for RTM-6 who provides a value of 41.12 MPa. A larger element size of 1.0 mm decreases the numerical effort significantly compared to a smaller element size of 0.5 mm. An element size of 1 mm and three elements across the length of the cohesive zone thus appear sufficient in combination with the material properties as summarized in Table 5.8.

Table 5.8 – summary of CFRP interface properties used for the cohesive zone model

l_{cz} [mm]	h_{cz} [mm]	K_N [MPa/mm]	K_T [MPa/mm]	G_{Ic} [J/m²]	G_{IIc} [J/m²]	τ_I^0 [MPa]	τ_{II}^0 [MPa]
3.23	0.001	$7.0*10^4$	$3.2*10^4$	426	1500	33	42

The ply by ply based model was also applied for simulation of the 3PB tests. For this purpose two modeling approach were compared. The first applies the ply by ply model for the full 3PB specimen of size 100 x 15 mm as shown in Figure 5.22. The second model uses instead a hybrid approach coupling the numerically expensive ply by ply model in the center with a simpler stacked shell laminate further outward as shown in Figure 5.23. The different sections are connected with a tied contact which transfers both node translations and rotations. Here the center section is of size 40 x 15 mm.

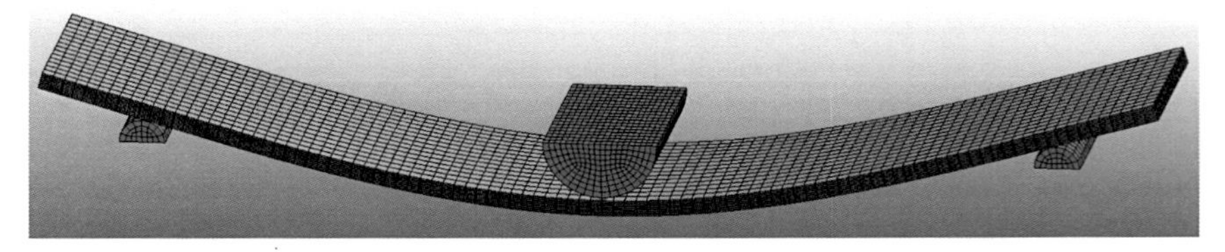

Figure 5.22 – simulation model of 3PB test with one shell element (grey) per ply connected by cohesive zone elements (blue)

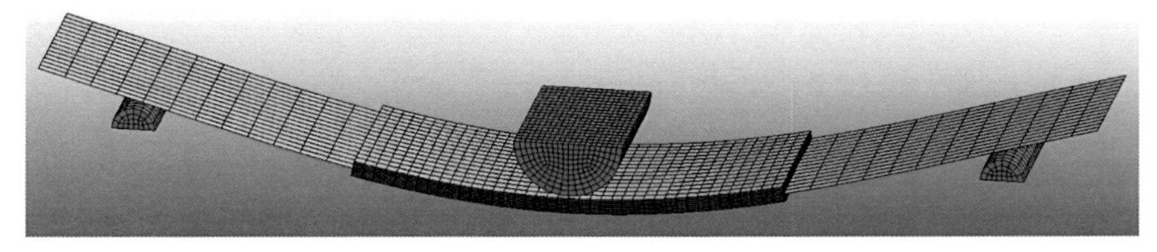

Figure 5.23 – simulation model of 3PB test using a hybrid approach with one shell element (grey) per ply and cohesive zone elements (blue) in the center; this is combined with a single stacked shell element (grey) for the full laminate further outward

A comparison of both models and the initial stacked shell model was performed with the progressive failure model MAT54 and the material properties of Table 5.4, Table 5.5 and Table 5.8. The force vs. displacement results are shown in Figure 5.24. From this it becomes clear that the stacked shell model has a noticeably different stiffness compared to the more complex ply-by-ply models. Both ply-by-ply models have the same stiffness but fail at slightly different load levels with no immediate explanation for this. Generally both models underestimate the fracture load. As the hybrid approach is equivalent to the full sandwich model, final calibration was performed with this modeling approach.

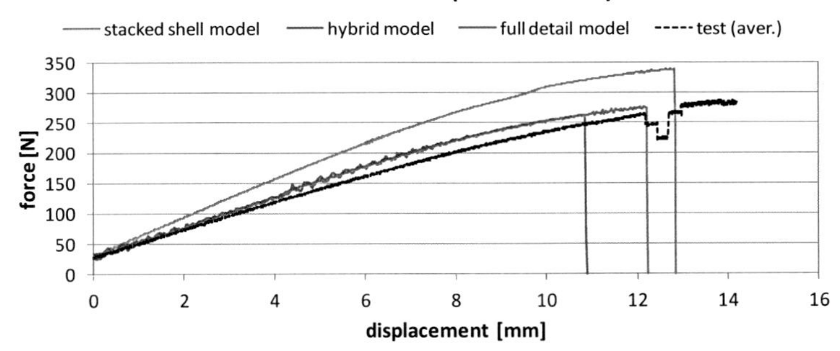

Figure 5.24 – simulation of 3PB test in laminate 0° orientation; comparison of hybrid model and full detail ply by ply model with test results

Initially the in-plane material properties of Table 5.4 and Table 5.5 were used for the simulation. Adjustment of material strength properties were required to account for greater stress concentrations in the ply-by-ply model. Failure of a single UD ply now leads to either a noticeable stiffness degradation or laminate failure due to immediate erosion of the affected element and related interlaminar crack growth. The element size of 1 mm is on the scale of fiber fracture and thus introduces fracture mechanics effects which the material model has to cope with despite – in case of MAT54 – being not specifically developed for this. Consequently the compressive failure strain was adjusted stepwise to 3.4% and the tensile failure strain to 2.8%. The simulation now correlate better with the experimental results as shown in Figure 5.25.

CFRP failure strains significantly larger than 2% may seem unrealistic. Here one has to keep in mind that this value describes only a very small material volume while material properties are usually provided for the bulk material. More importantly however the small element size leads to localization of damage. The affected area is limited thus increasing failure strains serves the purpose of smearing the damage across an area large enough to be critical for rupture. Fracture mechanics based models extend this idea in a more physical way as they apply the critical energy release rate and degrade the stiffness of the damaged material as damage accumulates. Here it should be kept in mind that particularly compressively loaded NCF composites can be subject to large damages before ultimate failure. Edgren et al. [Edg04] show that during compression after impact testing of NCF composites kink bands of 20 mm length may occur which is one order of magnitude larger than the applied element size of 1.0 mm and of the size of 3PB test specimen.

The 3PB tests were also simulated using the continuous damage model MAT262. Here a smaller element size of 0.6 mm had to be used for the fiber fracture cohesive zone. As the model includes a fracture mechanics based material law with damage, the fracture toughness values had to be estimated initially and were then adjusted according to the test results. Maimí, Camanho et al. [Mai07a] propose the use of the interlaminar fracture toughness G_{Ic} and G_{IIc} for the matrix tensile and respective compressive and shear failure modes. In addition the fracture toughness of fiber tensile and compressive failure are provided for the T300/913 prepreg material and were used as a starting value. The HTS fiber is about 20% stronger than the T300 fiber [Ten12, Tor12]. The influence of the different resin system and NCF textile cannot be assessed. Pinho et al. [Pin06b] report that one has to differentiate between the energy release rates obtained for crack initiation and growth. In case of T300/913 the initiation value was measured as 91.6 kJ/m² while the energy dissipation during crack growth is significantly higher with 133 kJ/m². The difference is explained by fiber bridging and pull-out in the wake of the crack tip. For compressive fiber fracture a crack initiation value of 79.9 kJ/m² is provided while the crack growth value of 143 kJ/m² is severely affected by secondary material failures aside from the crack tip such as matrix cracking and delaminations and thus questioned by Pinho et al. [Pin06b].

Simulation of the 3PB tests now revealed that in particular the initial guesses for the fracture toughness of the matrix dominated failure modes were too low and the properties were thus adjusted stepwise. There is no clear explanation to this but in the view of the author the effects of the NCF textile on matrix failure may contribute to this and should be subject of future work. Simulation and test results of 3PB tests in the laminate 0° orientation with both material models are provided in Figure 5.25.

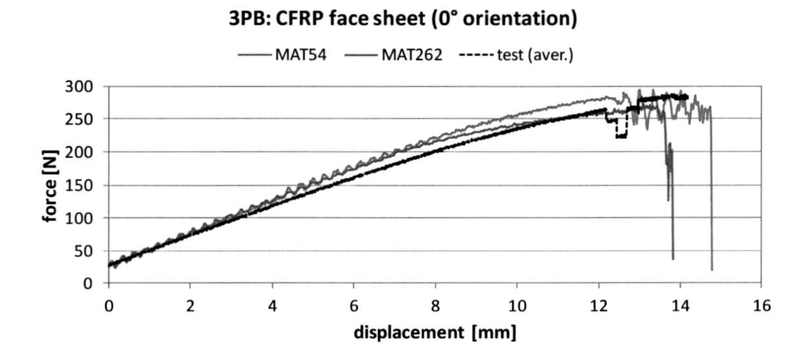

Figure 5.25 – simulation of 3PB test in laminate 0° orientation with different material models

Figure 5.25 reveals that both material models overestimate the bending stiffness of the undamaged test specimen slightly. Material damage however leads to stiffness degradation of the specimen during the 3PB test. MAT54 shows only minor degradation prior to specimen failure which is initiated during the simulation by tensile fiber fracture and subsequent element erosion. Differently MAT262 shows significantly more degradation prior to failure which highlights the effects of fracture mechanics connected with property degradation in this material model.

The same comparison of experiment and simulation was also performed for the 90° oriented 3PB specimens as shown in Figure 5.26. Generally the simulation results show similar trends as previously observed and are capable of modeling specimen stiffness and damage evolution. MAT262 however predicts material softening due to damage significantly better than MAT54. Furthermore MAT262 initiates specimen failure at 18 mm displacement while MAT54 is not capable of simulating specimen failure within the investigated maximum displacement of almost 20 mm.

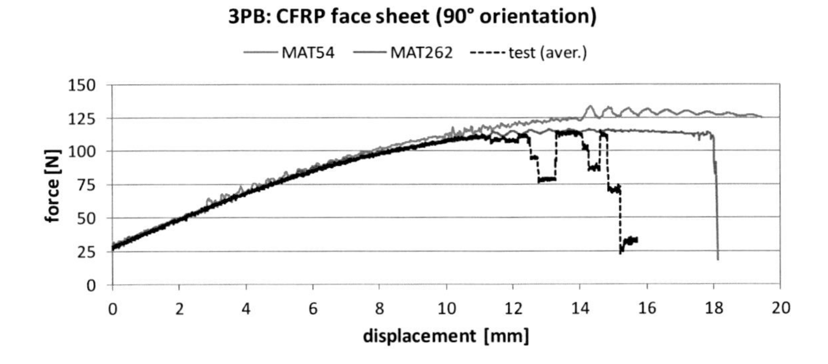

Figure 5.26 – simulation of 3PB test in laminate 90° orientation with different material models

Table 5.9, Table 5.10 and Table 5.11 summarize the final material properties used for the simulation of the NCF UD plies for modeling the CFRP face sheets using the more detailed ply-by-ply approach.

Table 5.9 – elastic properties of the NCF UD ply

ρ [g/cm³]	h_{ply} [mm]	E_{11} [MPa]	E_{22} [MPa]	G_{12} [MPa]	G_{23} [MPa]	ν_{12} [-]	ν_{23} [-]
1.51	0.132	126000	9100	4600	3800	0.28	0.2

Table 5.10 – strength properties of the NCF UD ply

X_T [MPa]	X_C [MPa]	Y_T [MPa]	Y_C [MPa]	S_L [MPa]	ε_{XT} [%]	ε_{XC} [%]	ε_Y [%]	ε_{SL} [%]	$\hat{\tau}_y$ [MPa]	G_{tan} [MPa]
2300	1400	80	200	125	2.8	3.4	20	20	100	180

Table 5.11 – fracture mechanics properties of the NCF UD ply

G_{XT} [kJ/m²]	G_{XT0} [kJ/m²]	X_{T0} [MPa]	G_{XC} [kJ/m²]	G_{XC0} [kJ/m²]	X_{C0} [MPa]	G_{YT} [kJ/m²]	G_{YC} [kJ/m²]	G_{SL} [kJ/m²]
100	25	200	90	20	150	2.1	7.5	7.5

5.3 Modeling of the foam core

5.3.1 Mechanical behavior and applied material model

Polymethacrylimide foam (PMI) – better known under the brand name Rohacell® – is when compared to other structural foams such as e.g. PUR, PET or PVC rather brittle but has superior stiffness and strength characteristics [Zen97]. Material properties of the Rohacell RIST PMI foam used for the sandwich core are available based on the manufacturer's data sheets [Evo13]. Table 5.12 summarizes the material properties for different RIST foam grades. This data set was then enlarged by the experimentally determined nonlinear compressive response, which describes complete failure through crushing of the foam cells.

Generally PMI foams are typically characterized as linear elastic until failure. A specific invariant based failure criterion tailored to the different compressive and tensile failure strength characteristics of the PMI foam has been developed by Kraatz [Kra07]. Kraatz investigated the failure behavior of Rohacell WF PMI foam under multiaxial loading conditions. This failure criterion has been applied to the RIST material by Rinker et al. [Rin08] for analyzing the results of four point bending tests. Application of the invariant failure criterion to the 71RIST foam used in this work is presented in appendix B.6.

Table 5.12 – material properties of Rohacell RIST foam core [Evo13]

	properties of Rohacell RIST PMI foam						
foam grade	ρ [kg/m³]	E_c [MPa]	$\hat{\sigma}_{c,C}$ [MPa]	$\hat{\sigma}_{c,T}$ [MPa]	G_c [MPa]	$\hat{\tau}_c$ [MPa]	$\varepsilon_{c,T}$ [%]
51RIST	52	75	0.8	1.6	24	0.8	3
71RIST	75	105	1.7	2.2	42	1.3	3
110RIST	110	180	3.6	3.7	70	2.4	3

One characteristic post failure behavior of closed cell foam materials is continuous compaction – also called crushing – of the cells in compression as described by Li et al. [Li00] for Rohacell 51WF. Foam crushing occurs in an elasto-plastic manner after initial compressive failure but without lateral contraction of the foam. Crushing continues until the foam cells collapse completely and full compaction is reached where the cell walls start blocking each other. Foam core crushing consumes a significant amount of energy and has thus to be taken into account when analyzing impact on sandwich structures.

Following an initial screening of applicable material models, the LS-DYNA materials 126 (MAT126) "Modified Honeycomb" and 142 (MAT142) "Transversely Anisotropic Crushable Foam" were selected for more detailed studies. Both models allow a separate treatment of normal and shear stresses based on a user defined post failure behavior and the possibility to apply strain rate sensitivity. Focus has been put on the correct description of core crushing and the failure behavior in tension and shear comparable to the criterion described by Kraatz [Kra07]. Furthermore the numerical implementation has to be stable and capable of dealing with material damage and large element distortions.

Material model MAT126 was initially developed for aluminum honeycomb in automotive crash applications but is also applicable to low density crushable foams with minimal lateral contraction. Fully anisotropic behavior can be defined. Normal and shear stresses are treated independent of each other based on a Poisson ratio of zero. Tensile, compressive and shear response are elasto-plastic without interaction of the yield stresses and based on user defined strain dependent yield stresses. This effectively includes a damage model as the yield stresses may also decline. Normal and shear strain cutoff levels can be specified for element erosion [Lst13a]. Two additional yield surfaces are also available for a more accurate description of anisotropic failure [Koj05].

Material model MAT142 was developed for the simulation of transversal isotropic crushable foams by Hirth et al. [Hir02]. The model behaves largely identical to MAT126 but is limited to transverse isotropic behavior. The normal and shear yield stresses are however coupled to each other using the Tsai-Wu criterion [Tsa72] for transversely isotropic materials. In consequence the elastic behavior of both material models is identical, the shape of the yield surfaces however not as the Tsai-Wu criterion describes an interaction of normal and shear stresses while MAT126 treats the stresses completely independent. Element erosion can be added to MAT142 as a complementary option (Mat_Add_Erosion).

Figure 5.27 illustrates the different yield surfaces in the normal-shear stress domain. The left chart compares the invariant failure criterion with MAT126 and MAT142 for Rohacell 71RIST using the material properties of Table 5.12. The right chart of Figure 5.27 displays for comparison the yield surfaces of Rohacell WF foams of different densities based on the invariant criterion of Kraatz as published by Roth and Kraatz [Rot08]. No similar chart has been published for the RIST material. It becomes clear that MAT142 is better capable to describe the material behavior of the Rohacell 71RIST foam.

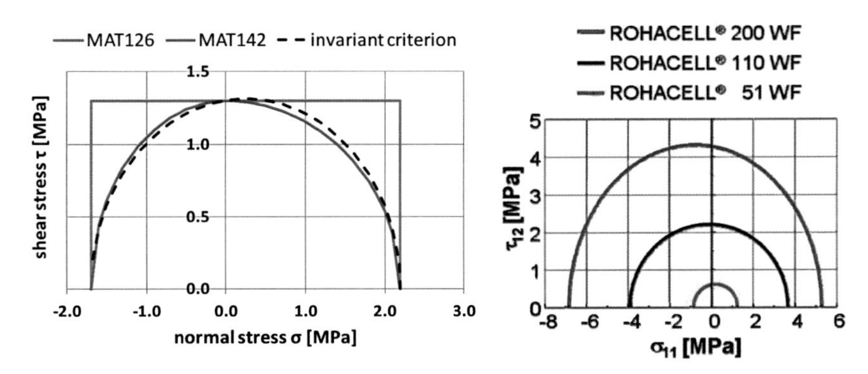

Figure 5.27 – yield surface of Rohacell 71RIST determined with the invariant failure criterion and LS-DYNA material models 126 and 142 (left); published yield surfaces of Rohacell WF foams (right) [Rot08]

Li et al. [Li00] report strain rate sensitivity of the Rohacell 51WF PMI foam properties. It is stated that the crushing stress is particularly sensitive to strain rates. This stress – which is on the order of the core compression yield stress $\widehat{\sigma}_{c,C}$ – increases from $\widehat{\sigma}_{c,cr} = 0.8$ MPa at a quasi-static strain rate of $\dot{\varepsilon}_{ref} = 10^{-3}$ s^{-1} to $\widehat{\sigma}_{c,cr} = 0.98$ MPa at a strain rate of $\dot{\varepsilon} = 1$ s^{-1}. This is an increase of roughly 20% at a still modest strain rate level. Li et al. [Li00] also state that the maximum shear strength is sensitive to strain rates as it drops by about 50% if the strain rates increases from $\dot{\varepsilon} = 10^{-1.4}$ s^{-1} to $\dot{\varepsilon} = 10^{-0.5}$ s^{-1}. The decrease of shear strength is however based on only two experiments and thus may be questioned. No results were provided on strain rate sensitivity of the stiffness.

Based on these test results Li et al. [Li00] provide a relationship for scaling the crushing stress of the Rohacell 51WF foam in the range of $\dot{\varepsilon}_{ref} = 10^{-3}$ s^{-1} to $\dot{\varepsilon} = 1$ s^{-1}. The crushing stress $\widehat{\sigma}_{c,cr}$ becomes

$$\widehat{\sigma}_{c,cr} = \widehat{\sigma}_{c,cr}^{ref} + C_{rate} \log\left(\frac{\dot{\varepsilon}}{\dot{\varepsilon}_{ref}}\right) \quad \text{with} \quad \widehat{\sigma}_{c,cr}^{ref} = 0.81 \text{ MPa} \quad \text{and} \quad C_{rate} = 0.056 \text{ MPa} . \tag{5.23}$$

As no test results were available for the Rohacell 71RIST foam, the scaling as described in equation (5.23) for the Rohacell 51WF foam was applied. The implementation of strain rates into MAT142 of LS-DYNA allows however only scaling of all yield stresses at once and not each separately. Thus the scaling used for the compressive response is applied to all strength properties of the foam core. It is noted that this will artificially increase the core shear strength which is critical to the failure mode core shear failure. As initiation of this failure mode is controlled by the rate insensitive cohesive elements in the foam core interface, the effect of rate dependent scaling of the core shear strength is considered as non critical. Final material properties of the foam core are summarized in appendix B.5.

5.3.2 Experiments

Two experiments were carried out to determine the mechanical response of the foam core material. The first experiment is out-of-plane crushing of the complete sandwich as shown in Figure 5.28. Here core thicknesses of 25.7 mm (test series M3) and 35.5 mm (test series M4) were compacted to 10% of their nominal core thickness at a speed of 0.01667 mm/s according to the standard DIN 53291. Figure 5.28 also shows exemplary one specimen before and after testing. It is clearly visible that the core volume reduces significantly with only minimal squeezing of the foam to the sides.

The experimentally determined load curve was then applied in the simulation model. Figure 5.29 shows the averaged test results together with the extrapolated load curve of the simulation. The core crushes very constantly after it passes a small peak stress at the onset of crushing. This peak is also included in the yield stress curve applied for the simulation. The measured crushing yield stresses average between 1.8 and 1.9 MPa which is roughly 10% higher than the stated value of the manufacturer [Evo13].

Figure 5.28 – test setup for in-plane compression (left), test specimens before and after testing (right)

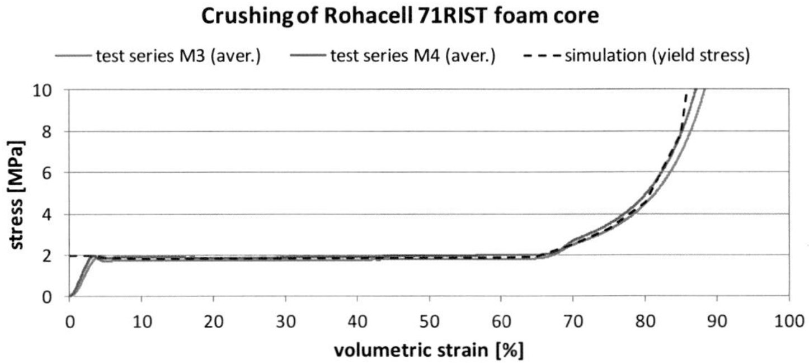

Crushing of Rohacell 71RIST foam core

—— test series M3 (aver.) —— test series M4 (aver.) - - - simulation (yield stress)

stress [MPa]: 0, 2, 4, 6, 8, 10

volumetric strain [%]: 0, 10, 20, 30, 40, 50, 60, 70, 80, 90, 100

Figure 5.29 – crushing behavior of Rohacell 71RIST foam

Foam core indentation was performed as the second calibration test. The aim of this test was to investigate the interaction of compressive and shear loads during core crushing and compression. Particularly foam core shear behavior is known to have strong localization effects resulting in core shear properties that are dependent on the individual test setup [Bat10, Cam12].

The core indentation test utilizes a hemispheric steel indenter of 25.4 mm diameter which was pushed into a sandwich specimen with the top face sheet removed. Figure 5.30 shows the test setup and a specimen after testing while Figure 5.31 displays the obtained test results. The indenter also has a shaft extension to 35 mm diameter shortly after the end of the hemispherical section. This extension is also pushed into the foam core resulting in the characteristic ring that can be seen in the tested specimen and the load increase noticed in the force vs. displacement plot.

The test results in Figure 5.31 show that the dominating energy consuming mechanism is core crushing. The load level corresponds well with the geometric volume displaced by the indenter as it first increases almost linearly, then turns to a nearly flat level until the shaft extension of the indenter is reached. Here the load level increases sharply but remains afterwards nearly flat again. Generally little scatter is experienced indicating stable foam properties.

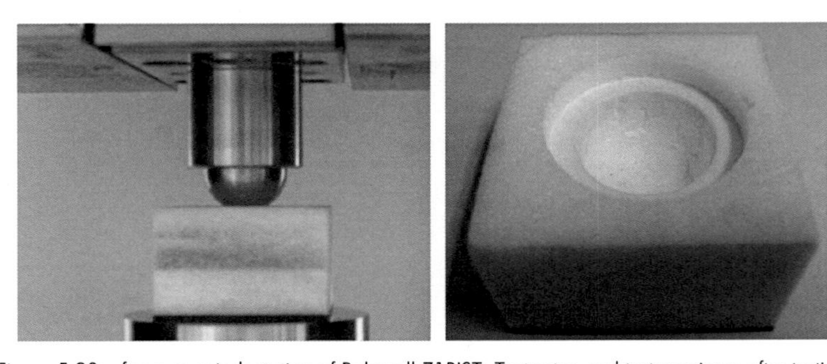

Figure 5.30 – foam core indentation of Rohacell 71RIST: Test setup and test specimen after testing

Foam core indentation of Rohacell 71RIST: Test results

test data --- average

force [N]: 0, 500, 1000, 1500, 2000, 2500, 3000

displacement [mm]: 0, 5, 10, 15, 20

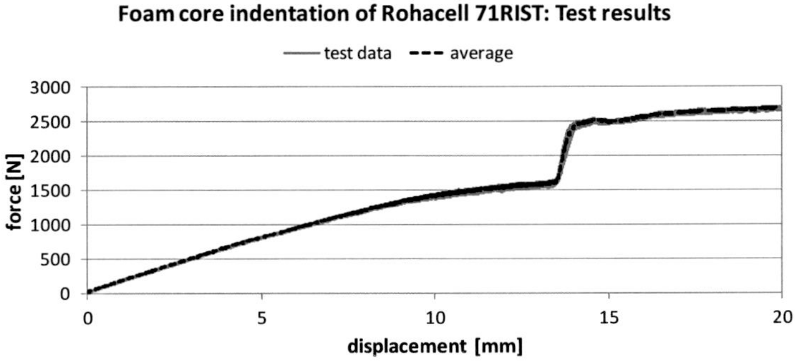

Figure 5.31 – foam core indentation of Rohacell 71RIST: Test results

This test was also simulated as shown in Figure 5.32. Initially it was not clear how the shear behavior of the core affects energy consumption in the test because core crushing is expected as the main energy consuming mechanism. Thus a number of different extreme cases were used as a baseline for comparison. Simulations have been performed both with MAT126 and MAT142 and the yield stress curve defined in the core crushing tests as shown in Figure 5.29.

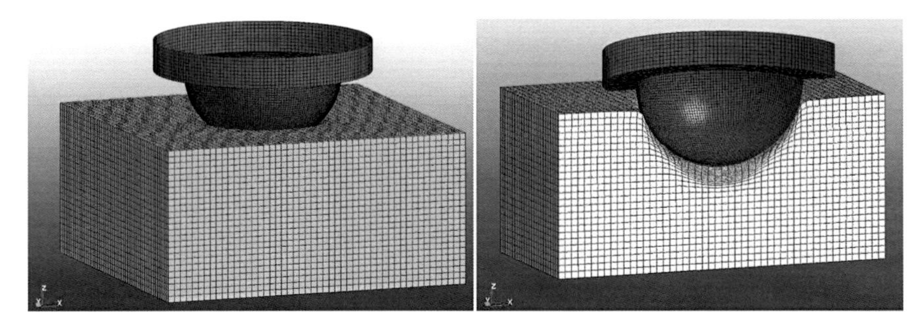

Figure 5.32 – simulation model of foam core indentation (left) and sectioning at 10 mm displacement (right)

Simulation and test results are shown in Figure 5.33 for MAT126 and Figure 5.34 for MAT142. Initially two extreme cases were investigated. The first simulation (blue) models the foam as elastic perfectly plastic until element erosion at a tensile failure strain of 5% takes place. As shear strains also have a tensile component, greater shear loads cause element erosion. In the second simulation (red) the foam is described as elasto-plastic, too but without element erosion.

In the first simulation (blue) significant element erosion was observed for both material models. Too many elements are thus eliminated too early due to the shear load and cannot consume energy via core crushing anymore. Thus both MAT142 and MAT126 underestimate the indenter force significantly. The second simulation (red) improves the behavior by omitting element erosion but leads to a load overestimation. MAT142 is better capable of describing the compression shear interaction as it overestimates the load only marginally starting at displacements between 5 – 10 mm while MAT126 overestimates the load more significant as shown in Figure 5.33 and Figure 5.34.

Physically crushing of the foam core damages the cell walls of the foam pores. This damage typically leads to localized material softening until lock up of the foam cells is reached and the compressive strength of the foam starts to rise again. As it was not clear how this affects the shear properties, different levels of eroding the material properties were tested numerically and compared to the results of the indentation test. Starting at a threshold value of 5% compressive volumetric strain the shear yield strength was degraded continuously until the compressive lock up strain of 65% was reached. From here on the shear yield strength was increased by the same scaling as the compressive response of the core.

The third simulation (green) presents the results of this shear strength degradation. Here the degradation of the shear properties was limited to 50% of the strength of the pristine material. Again MAT142 is noticeably better capable of describing the observed material response. Interestingly the effect of the shaft extension which penetrates as a sharp object into the core could be simulated reasonably well with MAT142 despite the large local deformations and related mesh distortion.

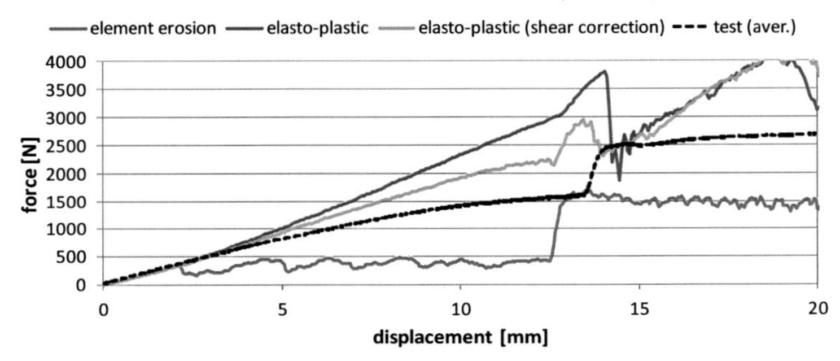

Figure 5.33 – simulation of foam core indentation tests of Rohacell 71RIST with MAT126; blue curve: Core crushing, elastic perfectly plastic behavior, element erosion at max. strain red curve: Core crushing, elastic perfectly plastic behavior, no element erosion green curve: Core crushing, elasto-plastic behavior with shear strength degradation

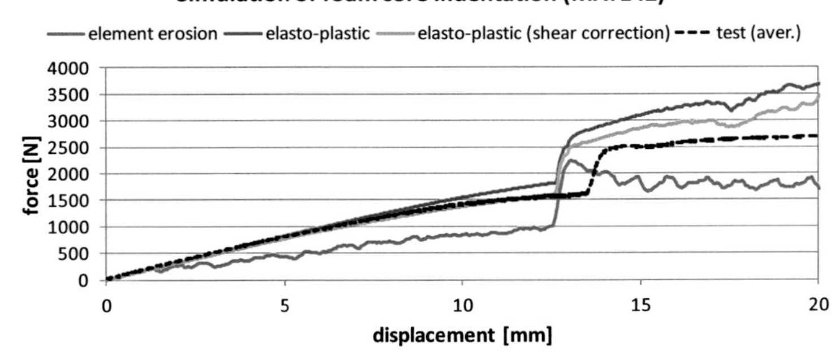

Figure 5.34 – simulation of foam core indentation tests of Rohacell 71RIST with MAT142; blue curve: Core crushing, elastic perfectly plastic behavior, element erosion at max. strain red curve: Core crushing, elastic perfectly plastic behavior, no element erosion green curve: Core crushing, elasto-plastic behavior with shear strength degradation

Core compression and indentation tests focus on energy consumption and the combined compression shear response. These tests replicate the type of core loading and damage that is initiated by an impact on a sandwich structure. Impact test results indicate that parts of the foam core are subject to significant shear loads due to out-of-plane forces on the sandwich where shear cracks may emerge. As element erosion cannot be used without sacrificing the core crushing response and related energy consumption, core shear failure in the simulation is limited to sandwich interface failure. Table 5.13 finally summarizes the material properties of the Rohacell 71RIST PMI foam applied in the simulation.

Table 5.13 – properties of the Rohacell 71RIST PMI foam core

ρ [kg/m³]	E_c [MPa]	G_c [MPa]	$\hat{\sigma}_{c,T}$ [MPa]	$\hat{\sigma}_{c,C}$ [MPa]	$\hat{\tau}_c$ [MPa]	ε_{lu} [%]	α_T [K^{-1}]
75	105	42	2.2	1.85	1.4	65	35 x 10^{-6}

5.4 Modeling of the sandwich structure

5.4.1 Interface crack growth

Sandwich structures compromise of three different constituents. These are the face sheets, the core and the interface between face sheet and core. A simulation strategy for the CFRP face sheets and the PMI foam core has been developed in the previous sections based on tests of the individual materials. This cannot be done with the interface as it only exists within the sandwich. Thus the next step is to develop a simulation model for the sandwich interface and validate it against test results.

An extensive investigation of interface crack growth in CFRP sandwich structures with Rohacell 71RIST foam core was performed by Rinker et al. [Rin11a] with results also published in [Joh09, Rin11]. Here the CFRP face sheets were manufactured of the same NCF composite and are thus fully representative of the test specimens used in this work. Single cantilever beam (SCB) tests were thus simulated for validation of the sandwich interface model. The simulation model was built with a fixed element edge length of 1 mm in the direction of crack growth and 5 mm in the other directions. A thermal load of $\Delta T = -150$ K is applied prior to the mechanical load with the model shown in Figure 5.35.

Results of three SCB tests were provided by Rinker [Rin10] for sandwich specimens with a nominal core thickness of 25 mm and a face sheet thickness of 2.25 mm but also published in [Rin11, Rin11a]. Figure 5.36 and Figure 5.37 display the test results and a comparison with simulation results for different values of G_{Ic}. Material properties for the face sheets and core were selected according to Table 5.9 and Table 5.12. The face sheet is modeled using a single shell element across the thickness.

The sandwich interface consists of resin filled foam cells that connect the face sheet with the core. The thickness of the interface can be estimated by the cell size of the core as only the outer layer of the PMI foam cells are open and thus fill with resin during manufacturing. As the cell size of the Rohacell 71RIST foam core is between 0.2 and 0.3 mm [Sae08,

Rin11a], the interface thickness was selected in the simulation to 0.2 mm. In the simulation the interface is modeled using cohesive zone elements similar to those applied for the interlaminar behavior of the CFRP face sheets.

Rinker et al. [Rin11a] determine the interface fracture toughness G_{Ic} to 190 J/m². Interface strength is limited by the foam core strength and was thus selected according to Table 5.13. Stiffness and thus compliance of the interface was selected based on the pure resin properties [Hex12] as the cohesive elements represent the thin layer of resin filled foam cells in the interface, which are otherwise not represented in the simulation model. The effect of the cohesive zone stiffness in the simulation model was however found to have little influence on the simulated SCB test results which may be attributed to the low stiffness of the core dominating local compliance. Finally the cohesive zone length and the number of elements across the cohesive zone were determined. Here the lower foam stiffness was used as it dominates local compliance and thus provides a conservative result. According to equations (5.20) and (5.21) the cohesive zone stretches across three to four elements for mode I and mode II crack opening which is sufficient for describing crack growth.

The simulation underestimates ultimate load and thus crack initiation but is well capable of describing crack growth. A possible explanation for this is that the simulation describes a perfect precrack while a teflon strip was inserted in the test prior to resin infusion [Rin11a]. This introduces a crack in the resin interface which then grows into the foam. This may also explain the lower stiffness of the simulation prior to crack growth observed in Figure 5.37.

It is noted that application of interface fracture toughnesses of 200 J/m² and 175 J/m² that are closer to the measured value of 190 J/m² overestimates the force required for crack growth. Use of the pure foam fracture toughness of 150 J/m² also reported by Rinker et al. [Rin11a] however describes the tested crack growth behavior best. This difference may be attributed to thermal pre-stressing of the foam core which introduces mode mixity and thus leads to an apparently different energy absorption in the sandwich test [Rin11a, Yok11]. As the simulation model however applies the pure foam value in combination with the thermal load, the interface response agrees between simulation and experiment best when using the pure foam value as shown in Figure 5.37.

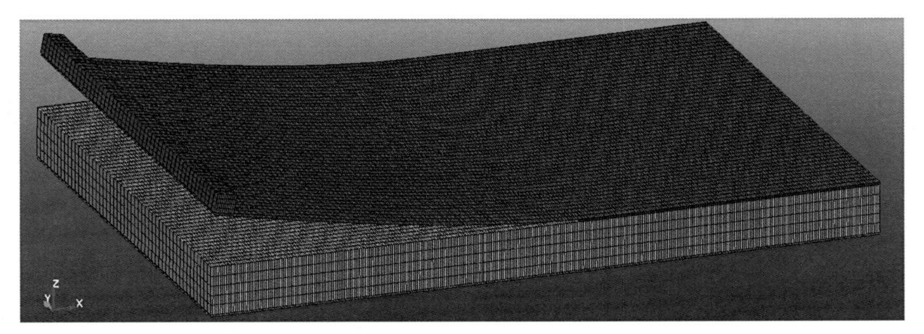

Figure 5.35 – simulation model of the SCB test with cohesive elements (red) for the sandwich interface

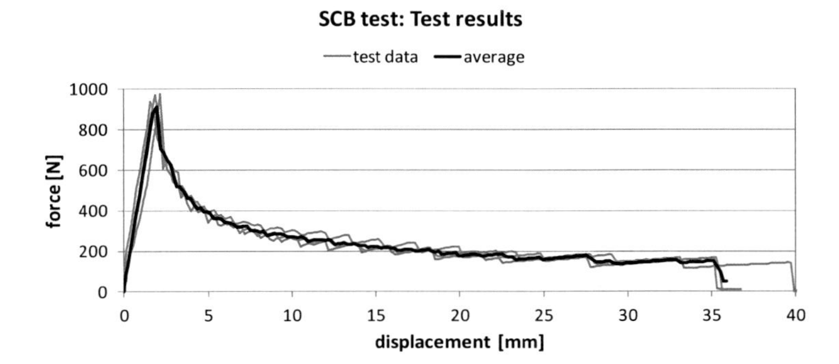

Figure 5.36 – SCB test with Rohacell 71RIST: Test results and average value

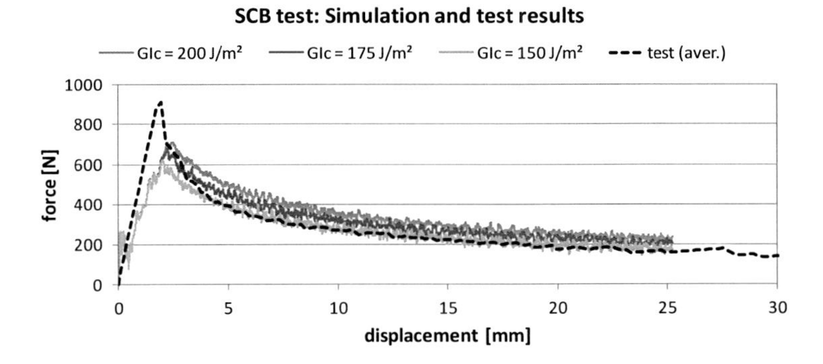

Figure 5.37 – SCB test with Rohacell 71RIST: Test and simulation results with different fracture toughness values

The SCB test investigates sandwich interface fracture mechanics in crack opening mode I while during the impact tests damage growth is primarily driven by crack opening mode II. The SCB test was however selected as the test results describe significant quasi-static crack growth. The cracked sandwich beam (CSB) test used by Rinker et al. [Rin11a] for investigation of crack opening mode II are limited to ultimate load measurements which then lead to unstable crack growth through the entire specimen. As ultimate load measurement are due to the artificial interface starter crack considered as rather imprecise, simulation and calibration of material properties based on the SCB test is preferred. Alternatively mixed mode bending (MMB) tests provide more reliable results of mode II crack propagation and should thus be considered for future studies.

During impact testing damage initiation is not affected by the crack growth parameters of the interface as an initial crack does not exist. Instead crack initiation occurs due to plastic deformation across a large foam volume which then creates a crack after maximum deformation is superseded. Thus the strength properties of the cohesive zone are selected according to the strength parameters of the pure foam. The effect of the crack growth parameters is instead limited to the mechanical response after foam core shear failure occurred and thus affects only damage size. Table 5.14 summaries the interface properties.

Table 5.14 – properties of the sandwich interface cohesive zone elements

h_{intf} [mm]	K_N [MPa/mm]	K_T [MPa/mm]	G_{Ic} [J/m²]	G_{IIc} [J/m²]	τ_I^0 [MPa]	τ_{II}^0 [MPa]
0.2	1.2*10⁴	0.45*10⁴	150	150	2.2	1.4

5.4.2 Static indentation of sandwich panels

The most complex test used for validation of the numerical model prior to the full impact test is a static indentation test of a sandwich plate. For this purpose six test specimens of size 100 x 100 mm were cut from a larger plate and statically indented with the same 25.4 mm indenter already used for the foam core indentation tests in section 5.3. The sandwich configuration used has a 25.7 mm thick sandwich core made of Rohacell 71RIST and 1.5 mm CFRP face sheets. The test setup is shown on the left side of Figure 5.38 while the right side displays a sectioned view of the numerical model shortly before face sheet rupture. The sectioned simulation model highlights three zones of the foam core using the colors yellow, brown and orange. In contrast all other figures of simulation models use for simplicity and clarity the color yellow for the complete foam core.

Figure 5.39 shows the results of the static indentation tests and their average. Generally scatter is very low until face sheet rupture. The first noticeable nonlinearity takes place at a load of about 1000 N. Shortly before this the linear elastic foam core behavior ends and the core starts to crush. In parallel delaminations emerge in the face sheet transforming the local response of the laminate from plate bending dominated into membrane dominated as previously described analytically in section 3.4. Core crushing and face sheet delaminations become visible in the simulation model shown in Figure 5.38 (right).

This continues until face sheet rupture takes place. The rupture load itself varies between the specimens from 5000 N to 6000 N and is followed by a noticeable sharp load drop. Rupture may take place stepwise as most load curves show two or three load drops. After face sheet rupture has taken place the load curve continues nearly stable between 4000 N and 5000 N until the maximum displacement of the test is reached.

The simulation model for this test has already the full complexity necessary for the impact tests. The model size is however significantly smaller due to the specimen dimensions of 100 x 100 mm. Furthermore the model is due to the lack of sandwich bending mechanically less complex than the impact model which makes this test ideal for calibrating contact laws of the different sandwich parts. Initially no thermal load is applied.

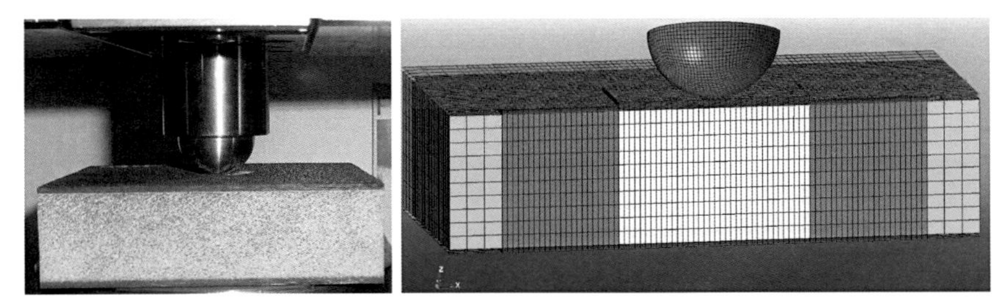

Figure 5.38 – sandwich static indentation: Test setup (left) and numerical simulation model (right), the sectioned simulation model highlights three foam core zones (yellow, brown and orange) and cohesive elements (red and light blue) at the upper and lower sandwich interfaces

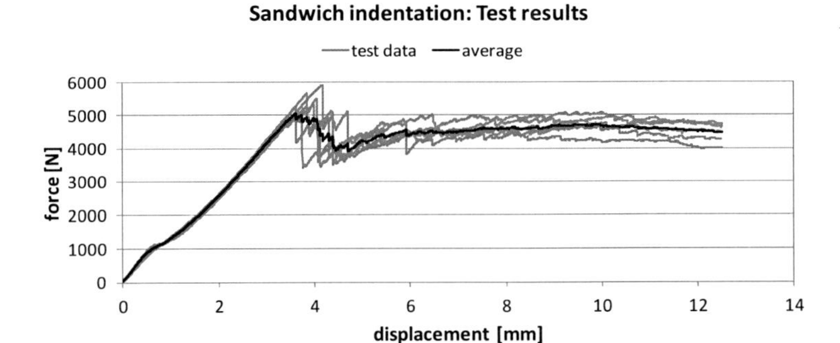

Figure 5.39 – sandwich static indentation: Test results

Considering the different discretization levels of sandwich structures previously discussed in section 5.1.2, a hybrid approach is applied. This approach composes of a detailed sandwich model in the impact region in the center of the specimen and a sandwich meso model that becomes stepwise less detailed to the outside. The indenter is modeled as a rigid half sphere with shell elements. Figure 5.38 (right) shows the simulation of the sandwich indentation tests just prior to face sheet rupture. Additionally Figure 5.40 and Figure 5.41 show sandwich indentation models after pull out of the indenter using either material MAT54 or MAT262 for the face sheets. Both simulation models are otherwise identical but the different face sheet material models and the mesh size in the center. Here MAT262 requires a finer discretization of 0.6 mm for a proper representation of the fracture mechanics while MAT54 uses a more coarse size of 1.0 mm as validated by the 3PB tests.

In both models the foam core is divided into three zones of which the central zone is immediately affected by the indenter. Here the upper face sheet is modeled using one shell element per ply and cohesive zone elements that connect the individual plies. The lowest ply is connected to the foam core with a cohesive zone element that represents the core interface. The core is modeled with solid elements and MAT142. The lower face sheet and its interface are modeled using a single shell element across the thickness and a cohesive element that connects the face sheet to the core.

In the intermediate zone the upper face sheet is modeled with a single shell element across the thickness. Mesh size increases moderately as shown in Figure 5.41. The face sheet connects to the foam core with cohesive elements that represent the core interface. A tied contact is applied for joining the upper face sheet in the center and intermediate zones as previously tested in section 5.2.5 for the simulation of 3PB tests. The foam core, lower face sheet and interface continue from the center.

The function of the outer zone in the model is limited to applying the correct boundary conditions to the sandwich structure. Thus cohesive zone elements are not anymore applied. The connection between the outer and intermediate zones is realized via shared nodes while the face sheet interface is realized with a tied contact.

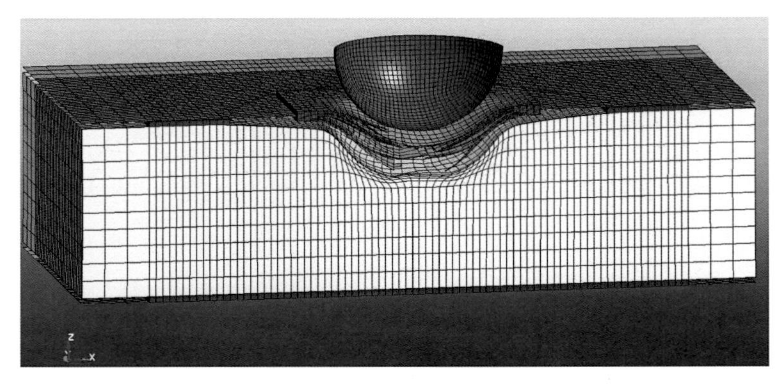

Figure 5.40 – sandwich static indentation after indenter pull out:
Simulation model with MAT54 applied for the face sheets

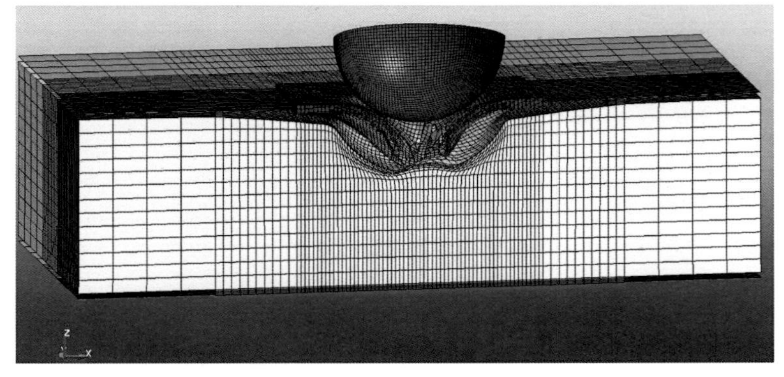

Figure 5.41 – sandwich static indentation after indenter pull out:
Simulation model with MAT262 applied for the face sheets

Comparing the sectioned views of the simulated indentation tests in Figure 5.40 and Figure 5.41, a few observations can be made. The size of core indentation and interface failure is comparable for both models but MAT262 initiates greater face sheet spring back compared to MAT54 which alters the appearance of the face sheet damage. Face sheet damage otherwise consists of fiber failure leading to element erosion and thus face sheet rupture and locally significant delaminations.

As the peak impactor force creates the largest sandwich bending moment and thus the greatest core shear stress during an impact, the description of the mechanical response up to face sheet rupture is important for model validation. The averaged test curve in Figure 5.39 indicates that face sheet failure takes place at about 5000 N. This underestimates the failure load slightly as most specimens fail at around 5500 N but slightly different displacement levels. The lower value of the averaged curve is thus an effect of the averaging process itself.

Figure 5.42 compares the simulated force vs. displacement curve with the test results. Here both material models overestimate the initial stiffness slightly but otherwise describe the initial face sheet response well. MAT54 (blue curve in Figure 5.42) initiates first ply failure and then face sheet rupture slightly too early at around 4500 N and 5000 N respectively. After face sheet rupture the indenter force drops stepwise to around 3000 N but then picks up again and reaches a plateau load of about 4500 N. In contrast MAT262 (red curve in Figure 5.42) initiates first ply failure already at around 4000 N but overestimates the face sheet rupture with a peak force of 6000 N. It then underestimates the later plateau load with only 4000 N slightly.

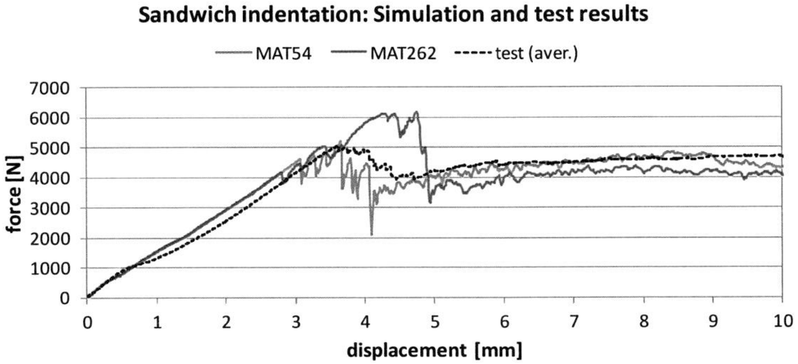

Figure 5.42 – sandwich static indentation: Simulation and test results

5.4.3 Thermal loads and strain rates

Besides the mechanical loading of the sandwich both thermal stresses and strain rate sensitivity may affect the sandwich impact response. The sandwich structure – particularly the foam core – is loaded by thermal strains which date back to manufacturing as explained in section 5.1.3. Strain rates have to be analyzed to check if rate dependent material properties of the CFRP and the foam core have to be considered. The sandwich indentation test is suited well for analyzing both effects.

The effect of thermal loads was investigated by comparing four different simulations with ΔT = 0 K, -75 K, -150 K and -225 K applied. Figure 5.43 shows the force vs. displacement plots with MAT54 applied and Figure 5.44 those with MAT262 applied. The different simulation results were artificially shifted for better distinction of the individual curves. The results of all simulations are much alike until first ply failure occurs but a noticeable oscillation. This oscillation is initiated by the fast application of the thermal load prior to indenter movement and results in a vibration of the foam core at its natural frequency. This oscillation is thus interpreted as an artificial effect of the simulation.

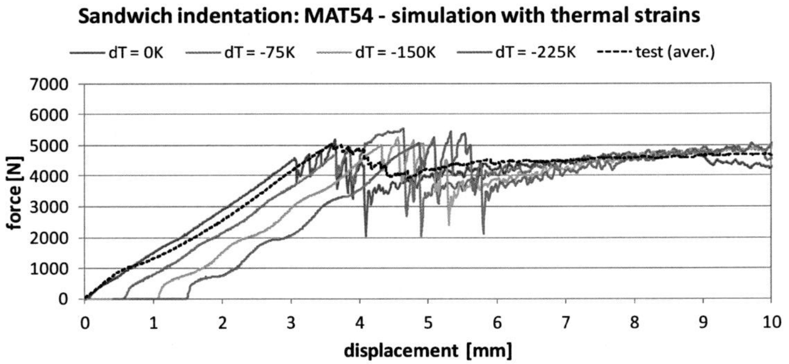

Figure 5.43 – sandwich indentation: Effect of thermal loads using MAT54

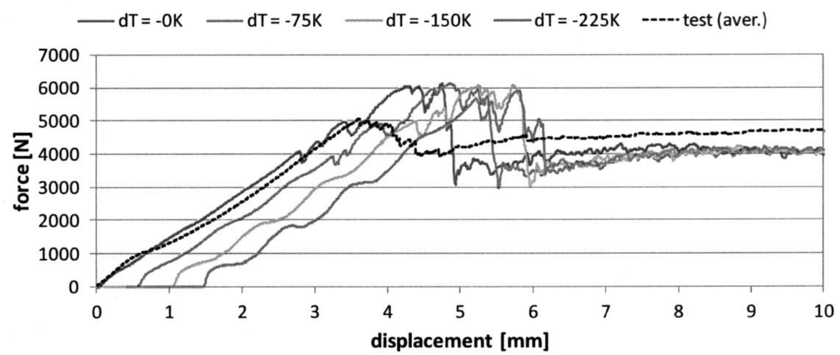

Figure 5.44 – sandwich indentation: Effect of thermal loads using MAT262

Summarizing the simulation results of Figure 5.43 and Figure 5.44 only minor effects of the thermal load were observed. Starting with Figure 5.43 and thus MAT54, the peak load that initiates with face sheet rupture varies between 5000 N and 5500 N for different ΔT. For ΔT = 0 K first ply failure takes place at around 4500°N but increases then to 5000 N for ΔT = -150 K and -225 K. In case of Figure 5.44 and thus MAT262 the effect of thermal stresses are similar as first ply failure shift with a rising thermal load from about 4000 N to about 5500 N for ΔT = -225 K. The peak load corresponding with face sheet rupture stays with a value of 6000 N nearly constant. Thus thermal loads lead in the simulation model to a delay of first ply failure which however has only minor effects on the peak load corresponding with face sheet rupture.

For investigation of strain rate effects core and face sheets were treated separately. As the experimental indentation tests were conducted sufficiently slow to avoid rate effects, the results of these tests are used as a static reference and compared to simulation results at higher velocities with and without strain rate effects considered. In the simulation the tests were speed up from 0.0333 mm/s to 5000 mm/s indenter velocity. Even though this velocity increase is conducted only for reducing the calculation time, an impact velocity of 5000 mm/s is typical for low velocity impact and thus serves well to estimate the effect of rate dependent material properties on the sandwich indentation and impact responses.

Figure 5.45 shows force vs. displacement of the indenter for four different simulations and the static test results as reference. The simulations are performed with a thermal load of ΔT = -150 K applied. The reference simulation (blue curve) does not include rate dependent material properties. The other simulations include rate dependent strength properties applied in the CFRP face sheets only (red curve), in the foam core only (orange curve) and in both (gray curve). As MAT262 does not allow application of rate effects, the investigation of strain rate effects was limited to the foam core material. The applied material properties are summarized in appendix B.7.

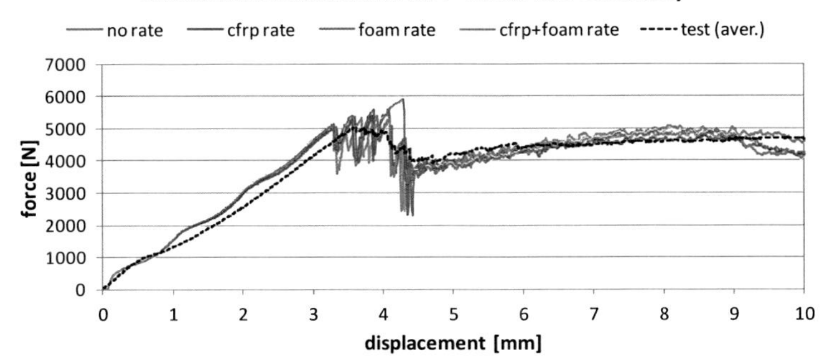

Figure 5.45 – effect of rate dependent strength properties on sandwich indentation with MAT54; no rate effects considered (blue), rate effects in the face sheet only (red), rate effects in the foam core only (orange) and rate effects considered in both (gray)

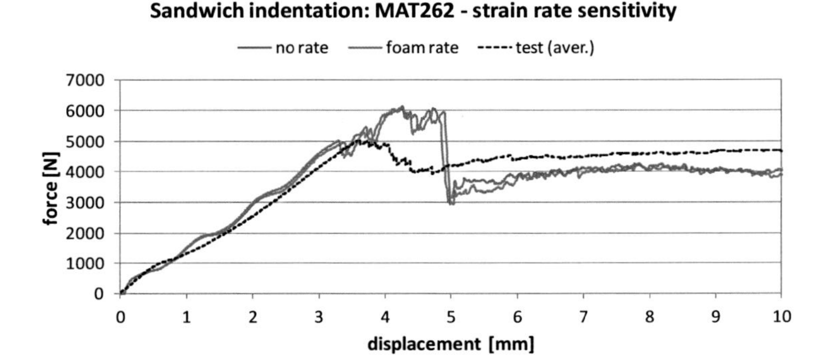

Figure 5.46 – effect of rate dependent strength properties on sandwich indentation with MAT262; no rate effects considered (blue) and rate effects in the foam core only (orange)

Referring to Figure 5.46 the addition of rate effects to the core stiffens the sandwich impact response slightly and leads to slightly earlier failure of the face sheet at nearly the same load level. The addition of rate effects to the face sheet has a very limited influence on face sheet rupture but leads to a modest decrease of the crushing load by about 200 N as visible in Figure 5.45. Combining rate effects in the core and the face sheets finally results in case of MAT54 in an overall slightly stiffer and stronger simulated indentation response compared to the model without rate effects.

It is thus concluded that rate sensitivity of the face sheet and core strength have only a limited influence on the mechanical response of the sandwich at 5000 mm/s indenter velocity. As the numerical cost of their implementation is however small, rate effects were applied in the full impact model in addition to the thermal reference load of ΔT = -150 K.

5.5 Simulation of low velocity impact

The numerical model for simulating low velocity impact on sandwich structures applies the same modeling approach as the earlier described model for static indentation. Cross-sectional views of the simulation model are provided in Figure 5.47 and Figure 5.48. The different zones in the center of the specimen are identical to the previous indentation model while the outer zones are extended and adapted to include larger areas of cohesive zone elements in the sandwich interface where necessary.

Model size thus increases as the dimensions of the sandwich plate are 350 x 400 mm and thus significantly larger than the indentation model. The additional model size concentrates on the intermediate area between the fine mesh in the center and the coarse outside mesh. In this intermediate area a relatively fine mesh discretization has to be applied in the in-plane direction as rear face sheet delamination may take place while the out-of-plane direction may be discretized more coarse. For application of the boundary conditions the face sheets are clamped in the area of the picture frame of 250 x 300 mm size comparable to the impact tests. Also the thermal reference load of $\Delta T = -150$ K was applied.

Figure 5.47 – sectioned overview of the sandwich impact model: Impactor (green), face sheet (grey), face sheet interlaminar cohesive elements (blue), foam core (yellow) and sandwich interface (red)

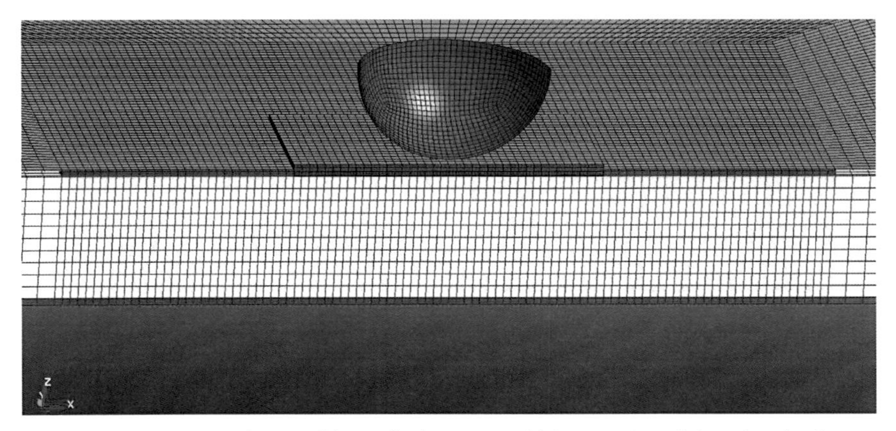

Figure 5.48 – sectioned center of the sandwich impact model: Impactor (green), face sheet (grey), face sheet interlaminar cohesive elements (blue), foam core (yellow) and sandwich interface (red)

5.5.1 Impact response

Five different sandwich configurations tested during the RT test series where simulated for experimental validation of the developed simulation model. All simulations were performed both with MAT54 and MAT262. The selected specimens generally represent the investigated range of sandwich configurations and were made of 1.5 mm (RT_10.0, RT_16.3 and RT_25.7) or 2.25 mm thick face sheets (RT_H2a and RT_H2b). Core thickness varies from 10 mm to 25.7 mm. Of these two configurations were subject to foam core shear failure (RT_10.0, RT_H2a) and three to face sheet rupture (RT_16.3, RT_25.7, RT_H2b) – see also Figure 4.27. Additionally configuration RT_10.0 was simulated with three different impact energies of 12 J, 20 J and 35 J to investigate the predicted damage size due to core shear failure.

Starting with the specimens with 1.5 mm thick face sheets configuration RT_10.0 was initially simulated with 20 J impact energy sufficient for the creation of core shear failure but not for face sheet rupture. The configurations RT_16.3 and RT_25.7 were instead simulated with 35 J impact energy to initiate face sheet rupture as observed during the tests. For comparison with the impact test results Figure 5.49 shows the force history of simulation and test while Figure 5.50 shows sectioned simulation models with MAT54 and MAT262.

The impact response in Figure 5.49 reveals reasonable agreement of simulation and test. The peak or threshold force that initiates core shear failure and the post failure force level are slightly overestimated. Besides that the impact response is generally captured well. The choice of material model for the CFRP face sheets does not make a noticeable difference which is attributed to the fact that only minor face sheet damage occurred during the impact. The sectioned views also show that both models describe the creation of core shear failure correctly. Damage size is moderately overestimated as shown in Figure 5.57.

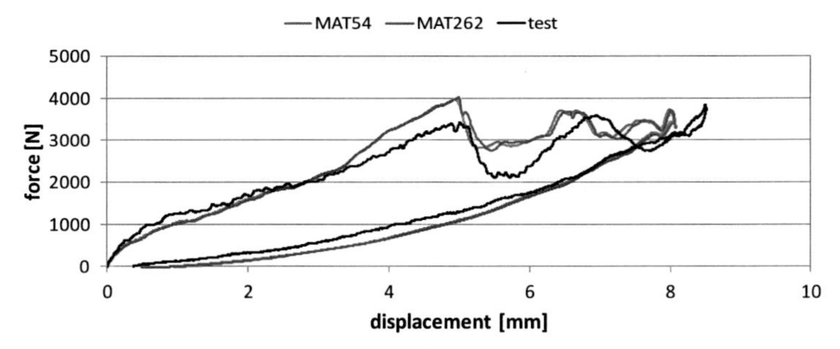

Figure 5.49 – force history of 20°J impact test and simulation of specimen RT_10.0_P1

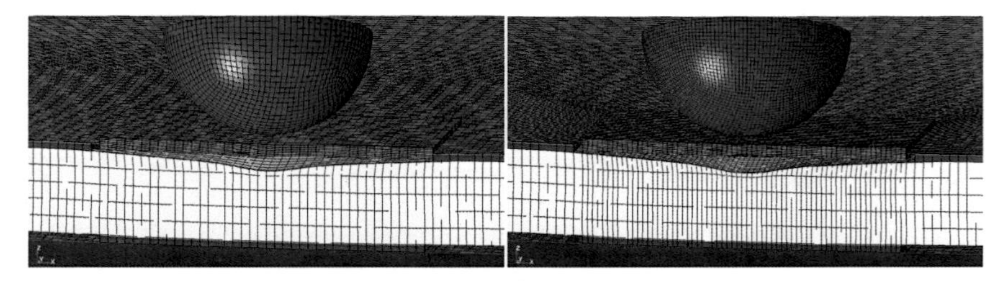

Figure 5.50 – sectioned simulation models after 20 J impact on specimen RT_10.0_P1 with 1.5 mm face sheets and 10 mm core using MAT54 (left) and MAT262 (right)

Concentrating on the failure mode face sheet rupture the force history of 35 J impact tests on specimens RT_16.3 and RT_25.7 is shown in Figure 5.51 and Figure 5.52 respectively. Both specimens failed during testing by face sheet rupture. Here MAT54 shows generally a better agreement with the test results as the force history follows the measured value very closely until the final part of the impact that describes the rebound of the impactor. MAT262 overestimates the face sheet rupture load moderately and then underestimates the post rupture plateau value also moderately but is significantly better capable to describe impactor rebound.

This difference becomes even more imminent as one compares sectioned views of the simulated sandwich impact damage shown in Figure 5.53. Here the face sheet contains – despite cracks – in those simulations with MAT54 significant plastic deformation and experiences only limited spring back. The impactor thus loses at the end of the impact process contact with the face sheet earlier than in reality. Differently MAT262 describes significant spring back of the damaged face sheet. The impactor thus stays longer in contact and experiences during face sheet spring back continuously a moderate force and thus acceleration until it is eventually pushed out of the damaged structure. This leads to the creation of a significant cavity below the damaged face sheet.

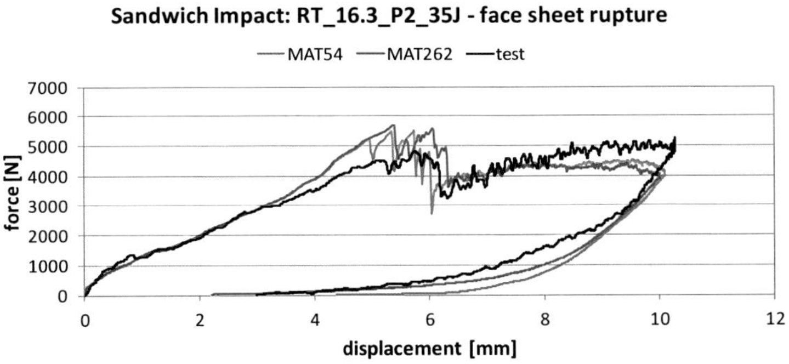

Figure 5.51 – force vs. displacement of 35°J impact test and simulations of specimen RT_16.3_P2

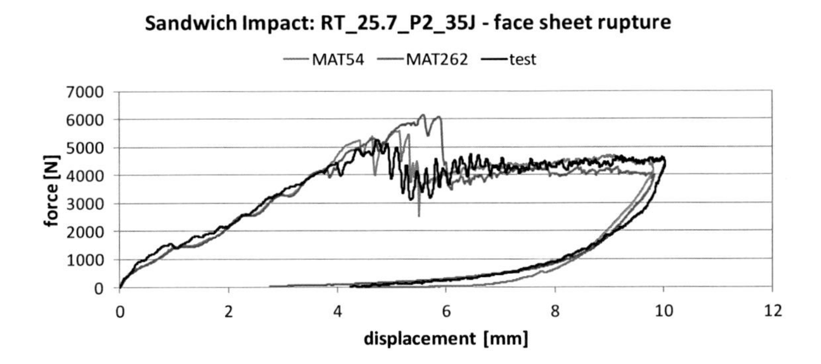

Figure 5.52 – force vs. displacement of 35°J impact test and simulation of specimen RT_25.7_P2

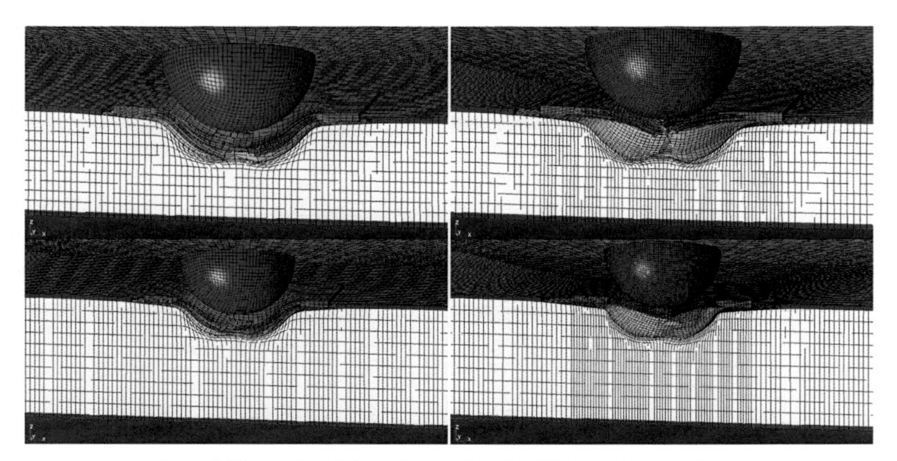

Figure 5.53 – sectioned simulation models after 35 J impact on specimens RT_16.3_P2 (upper row) and RT_25.7_P2 (lower row) using MAT54 (left) and MAT262 (right)

Additionally tests on configurations RT_H2a and RT_H2b each with a 50 J impact were selected to analyze the impact response of a sandwich configuration with greater energy and face sheet thickness. The impact energy was increased accordingly to 50 J. The configurations have 2.25 mm thick face sheets and core thicknesses of 16.3 mm and 25.7 mm. Force vs. displacement of test and simulation of the specimens RT_H2a_P2 and RT_H2b_P2 are compared in Figure 5.54 and Figure 5.55 respectively. Additionally sectioned views of the simulation model are shown in Figure 5.56.

Testing in chapter 4.2 showed that specimen RT_H2a_P2 was subject to core shear failure while specimen RT_H2b_P2 experienced face sheet rupture. The experimental force history of both specimens thus differ noticeably as after surpassing the peak load of around 8 kN the impactor force of specimen RT_H2a_P2 drops to a range of 4 to 5 kN due to the large loss of bending stiffness while for specimen RT_H2b_P2 it stays around 6 kN. The greater load of specimen RT_H2b_P2 can be correlated with continuous face sheet rupture and core crushing, which did not take place in the thinner specimen RT_H2a_P2 due to core shear failure.

The simulations of both specimens with MAT54 and MAT262 describe the sandwich failure mode correctly. The simulated force history of specimen RT_H2a_P2 however deviates noticeably as the simulated load curve overestimates the peak and post failure loads independently of the applied face sheet material model. There is no immediate explanation for this load overestimation. It is however noted that a quantitatively smaller but qualitatively similar overestimation of the core shear failure and post failure forces was also observed in the simulation of specimen RT_10.0_P1. In contrast the force history of specimen RT_H2b_P2 with face sheet rupture is captured well by the simulation independently of the use of MAT54 or MAT262 for the face sheet.

Referring to Figure 5.56 the shear failure of specimen RT_H2a_P2 becomes visible in the simulation by erosion of the interface cohesive elements at the rear face sheet. Directly at the point of impact the interface remained intact. Only some layers of the impacted face sheet failed its rear side. Finally a cavity and core debonding formed in the area with core crushing due to spring back of the damaged face sheet.

Specimen RT_H2b_P2 was subject to face sheet rupture in experiment and simulation. The interface to the rear face sheet thus remained intact with no damaged cohesive zone elements observed. Instead failure occurred at the impacted face sheet which failed by rupture. Spring back of the damaged face sheet was small in the simulation when using MAT54 and significantly greater when using MAT262 which corresponds with the more gradual rebound of the impactor force in Figure 5.55. Core damage on the impacted sandwich side can be distinguished further into damage size and depth of the crushed core zone. As expected face sheet rupture leads to a more concentrated core crushing while damage below an intact face sheet has the shape of a local depression.

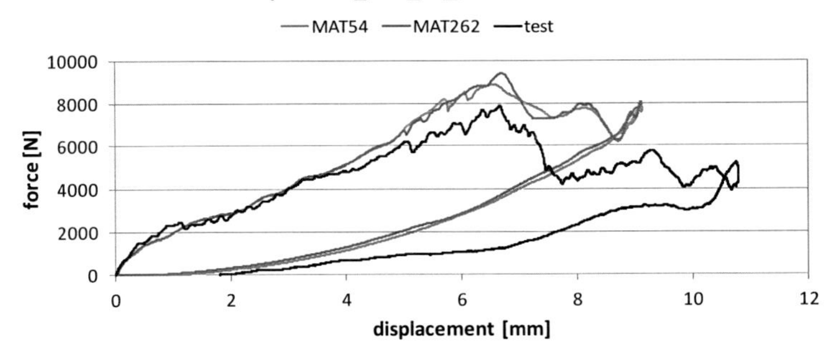

Figure 5.54 – force vs. displacement of 50°J impact test and simulation of specimen RT_H2a_P2

Figure 5.55 – force vs. displacement of 50°J impact test and simulation of specimen RT_H2b_P2

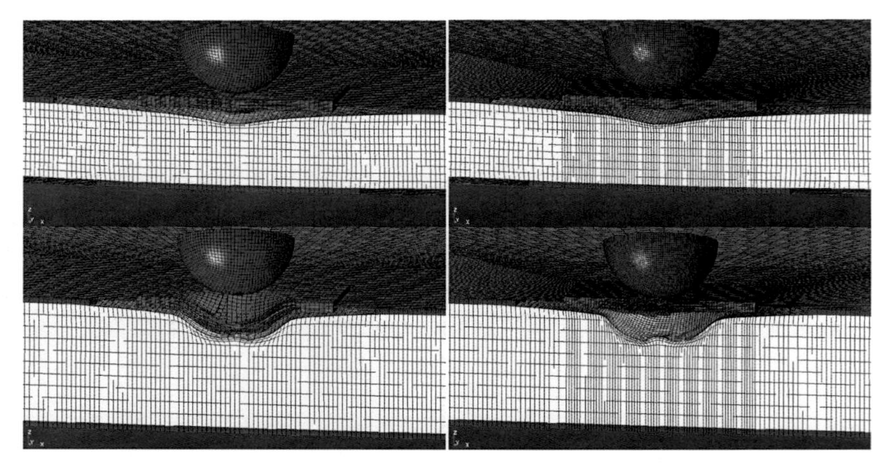

Figure 5.56 – sectioned simulation models after 50 J impacts on specimens RT_H2a_P2 (upper row) and RT_H2b_P2 (lower row) using MAT54 (left) and MAT262 (right)

5.5.2 Damage size and impact failure mode

Damage size in the sandwich specimens must be characterized depending on the failure mode as shown previously for the impact test series RT in Figure 4.9. Here core shear failure emerged as the more critical failure mode as it initiates significantly greater failure sizes. In the simulation damage size of this failure mode is determined by the rear face sheet debond which is compared in Figure 5.57 for 12 J, 20 J and 35 J impacts on sandwich configurations RT_10.0 together with NDI images of the impacted specimens. Damage sizes generally agrees well except for the 12 J impact. During testing the 12 J impact energy was not sufficient for creating core shear failure. In contrast the simulated impact response is more sensitive with respect to core shear failure as the impact initiates a small rear face sheet debond.

For final verification of the quality of the model, the complete RT impact test series was simulated and the sandwich failure mode characterized. Figure 5.58 shows the observed failure modes of all simulated specimens using the previously applied failure mode map. Overall good agreement of the simulated failure mode is reached. It is however noted that the simulation underestimated the threshold energy for core shear failure for the configurations RT_10.0, RT_H2a and RT_H3a. The complete test results have already been depicted in Figure 4.27. Furthermore specimens RT_H1a_P2 and P3 – both subject to 35 J impact energy – were in the experiment subject to face sheet rupture and severe core crushing, which then led to crushing induced shear cracks and a local rear face sheet delamination. This behavior was not observed in the simulation and shows a limitation of the applied simulation model.

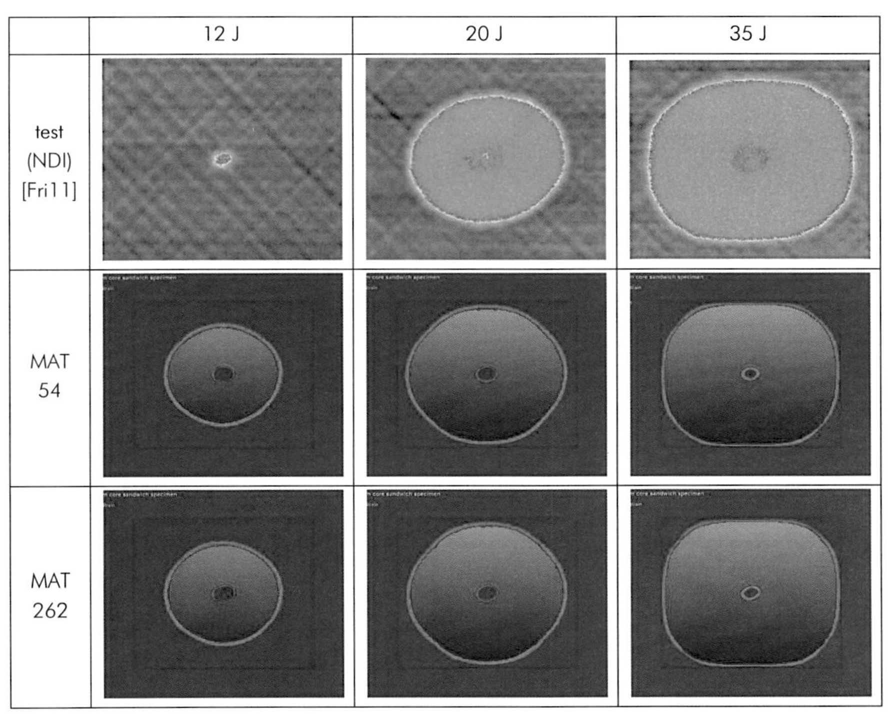

Figure 5.57 – impact damage of sandwich configuration RT_10.0: 12J (left), 20J (center) and 35J (right); upper row: NDI images; center and lower rows: Simulated rear interface damage with element erosion, use of material model MAT54 (center row) and MAT262 (lower row) for the CFRP face sheet; color code – red, green (high attenuation / damage) and blue (low attenuation / no damage)

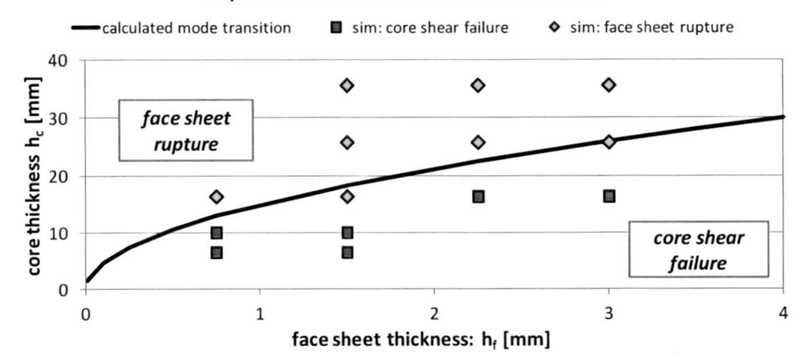

Figure 5.58 – impact failure mode map for low velocity impact of foam core composite sandwich plates with analytical solution and simulation results

5.6 Chapter summary

This chapter described the development of a simulation model for low velocity impact on foam core sandwich structures. As the simulation model has to be capable to describe different kinds of material damage resulting in the two governing sandwich failure modes face sheet rupture and core shear failure, relevant material laws and their respective properties were determined, calibrated and validated against test results. A hybrid modeling approach was selected for simulation of the sandwich structure that is composed of a sandwich meso model which describes core and face sheets separately in combination with a more detailed discretization of the impacted face sheet and sandwich interfaces where needed.

Following the building block approach tests and both analytical and numerical analysis have to be performed in parallel with stepwise increasing complexity. The intralaminar material properties of the NCF composite face sheets were initially determined using rules of mixture for the stiffness and literature results of equivalent unidirectional materials for the strength. These were then calibrated using three-point bending tests of the sandwich face sheets. As the intralaminar material behavior is highly complex, two different material models were investigated. The first was a simpler progressive failure model and the second a more complex continuous damage model with fracture mechanics based degradation of the material properties. The interlaminar properties of the face sheets were determined using literature results in consideration of the applied cohesive material law and were then calibrated against interlaminar shear tests.

Description of the foam core material was developed similarly with the elastic and initial strength properties based on material data sheets. This data was enlarged by a user defined core crushing response and the shear compression response calibrated against foam indentation tests. Comparison of two applicable material models showed that a crushable foam model based on an elasto-plastic response and a shear correction is well capable to describe the indentation response. The foam core interface was described using a fracture mechanics based cohesive material law which was validated against single cantilever beam tests of sandwich beams.

The full sandwich model was then validated using quasi-static sandwich indentation tests as reference. The indentation tests describe already the full complexity of the sandwich indentation response that leads to face sheet rupture. As the sandwich is not subject to any bending, face sheet rupture could be investigated separately of foam core shear failure.

Finally the developed simulation model was applied to low velocity impact. Simulation results were compared with test results previously presented in chapter 4. The simulation model was generally capable to describe the correct failure mode. In case of face sheet rupture agreement of simulation and test based on the force history was good. The progressive failure and continuous damage model were found to describe the impact response similarly well until spring back of the face sheet. Here the fracture mechanics based continuous damage model is noticeably better capable to describe spring back of the impacted face sheet including impactor rebound.

The force histories of the sandwich impact specimens that are subject to core shear failure agree moderately well with the test results. Initiation of failure is generally found to occur at slightly higher impact loads than those observed during the tests. The simulated post impact response is also moderately stiffer than the experimentally observed response. The choice of face sheet material model did not make a noticeable difference. Damage size of test and simulation agree reasonably well.

It is thus concluded that the developed simulation model is generally capable to describe the impact response and resulting failure mode reasonably well. As both material models describe for a reference specimen with 1.5 mm thick face sheets the failure mode face sheet rupture similarly well, future work on the analysis and simulation methods should concentrate on improving the foam core material model by e.g. enhancing the applied elasto-plastic crushable material model with fracture mechanics based material property degradation. Interaction of foam core crushing and core shear failure are here of primary interest.

The physically based continuous damage model has in comparison to the progressive failure model the advantage of relying completely on physical properties which will likely lead to less numerical calibration work but increased material testing on the coupon and element level. The improved material spring back response may not be of great importance for the description of the impact failure mode but indicates the improved quality of this type of material model. This comes however at significantly higher numerical costs due to the smaller element size required.

6 Sandwich failure mode parameters

6.1 Approach

Sandwich structures loaded by low velocity impact are subject to different failure modes as shown previously in section 3.5. The resulting impact damage is predominantly controlled by the occurrence of face sheet rupture and core shear failure as governing failure modes. From analytical criteria it is known that the shear stress in the core plays a critical role in determining the sandwich failure mode. If the shear load supersedes the foam core shear strength, the core material will deform and subsequently fail by cracks oriented perpendicular to the largest principal stress. The exact origin of crack initiation is unknown, but the largest and most critical section of crack growth occurs at the rear face sheet interface.

The finite element model developed in the previous chapter can describe the occurrence of both failure modes. The core is modeled as an ideally plastic material which may deform but not develop cracks. In the simulation crack initiation and growth is thus limited to the face sheet interfaces, which in case of the rear interface will fail due to a shear overload and subsequent crack growth. Thus using the available finite element model one can determine the failure mode of any particular sandwich configuration and loading. Additionally this model can be used to determine the maximum core shear load and compare it to the foam shear strength for calculation of a safety margin.

To investigate influencing parameters on the sandwich failure mode, the core shear stress was determined for different impact cases subject to changes of relevant impact parameters and boundary conditions. This allows quantifying the sensitivity of the specific case to foam core shear failure. Therefore a reference sandwich configuration is required, which is not subject to core shear failure as this would lead to interface failure and subsequent element erosion. Instead a sufficiently large safety margin to core shear failure is required which allows the measurement of a relevant increase of the core shear load in the elastic domain. The sandwich configuration RT_25.7 with 1.5 mm face sheets and a 25.7 mm core was selected based on these considerations. All specimens of this sandwich configuration failed by face sheet rupture when tested experimentally as shown in chapter 4.2. The maximum core shear stress $\bar{\tau}_c$ for configuration RT_25.7 is determined to 1.0 MPa using equation (3.39). Comparing this to the foam core shear strength of 1.4 MPa provides a safety margin of 40% which is considered adequate.

The reference impact energy was selected to 35 J as applied with specimen RT_25.7_P2. This impact energy was sufficient to initiate face sheet rupture. Core shear failure takes place in the cohesive elements of the rear face sheet interface. As cohesive elements are special purpose elements that do not describe the full stress state of the core, core shear stresses were measured in the core elements at the transition to the rear face sheet interface. This allowed the use of Tresca's shear stress criterion, which is not possible for the cohesive zone elements of the interface. The impact simulation of specimen RT_25.7_P2 provided with $\bar{\tau}_c = 1.046$ MPa a value slightly higher than the analytically determined load of 1.0 MPa. Comparing the impactor forces that initiate the core shear load reveals only minor differences. The analytical failure criterion of equation (3.38) estimates a face sheet rupture force of $\mathbf{F_{rup}} = 5462$ N, while the simulation determines an almost equal maximum impactor force of $\mathbf{F_{imp}} = 5553$ N as shown in the force vs. displacement plot in Figure 6.1. The simulation was performed using MAT54 for the face sheets and applies a thermal preload of -150 K. The analytical core shear failure criterion of equation (3.39) does instead not account for thermal stresses.

Figure 6.2 displays the core shear stress distribution of a section of 100 x 100 mm in the center of the sandwich that is measured at the rear face sheet interface. From this it becomes clear that shear stresses greater than 0.9 MPa concentrate in a circular area around the impact. The peak shear stress of 1.046 MPa is located perpendicular to the 0° face sheet orientation.

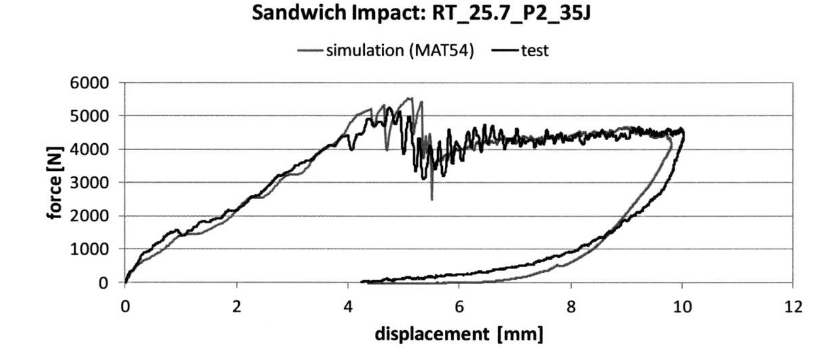

Figure 6.1 – force vs. displacement of test and simulation of specimen RT_25.7_P2 subject to a 35 J impact

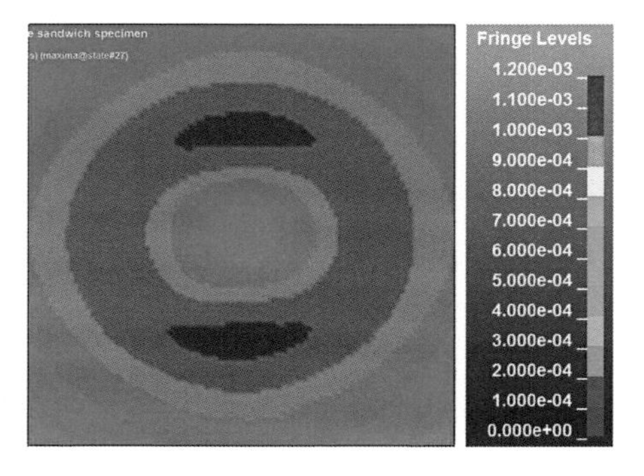

Figure 6.2 – maximum core shear stress in GPa at the rear face sheet interface during simulation of a 35 J impact on specimen RT_25.7_P2 with -150 K thermal load applied; area shown is 100 x 100 mm with a centrally placed impact

6.2 Thermal loads

Thermal loads in sandwich structures originate due to a mismatch of thermal expansion of face sheet and core. The effect of this on sandwich beam failure has been discussed by Rinker et al. [Rin08a]. Impact tests at -55°C, which were described in chapter 4.3, revealed that low temperatures have a noticeable effect on the sandwich failure mode. From the test results it is however not possible to differentiate the proportions of thermal strains and temperature dependent material properties on this effect. The simulation applies thermal strains and material properties separately and may thus distinguish between them.

Therefore impact simulations were performed with different thermal loads but constant material properties based on specimen RT_25.7_P2. The thermal load ΔT was applied prior to the impact and varied from 0 K to -225 K in steps of 25 K. Figure 6.3 shows the force vs. displacement curve of four different simulations each spaced apart by 75 K thermal load while Figure 6.4 plots the maximum shear load. As the peak impactor force varies noticeably with the thermal load which in turn affects the core shear load directly, both the maximum shear stress $\bar{\tau}_c$ in the simulation and the shear stress sensitivity parameter S_τ are plotted. This parameter is defined by the quotient

$$S_\tau = \frac{F_{imp}}{\bar{\tau}_c} \,. \tag{6.1}$$

S_τ describes the core shear stress relative to the maximum force of the impactor. Referring to Figure 6.4 it is noticed that the shear stress sensitivity increases nearly linear with the applied thermal load. This may be explained by superposition of thermal and mechanical loads. To reduce scatter a best-fit line is used in Figure 6.4. The total increase of the shear load remains however modest. At $\Delta T = 0$ K the best-fit value of S_τ is 0.165 MPa/kN which increases by about 18% to 0.195 MPa/kN for $\Delta T = -225$ K. The measured values of S_τ is 0.175 MPa/kN at $\Delta T = 0$ K and 0.197 MPa/kN for $\Delta T = -225$ K. The maximum shear stress $\bar{\tau}_c$ in the simulation experiences a similar trend as S_τ.

It is also noticed that thermal strains influence the impact response and the total core shear stresses. The simulated response shows with increasing thermal load a noticeable load drop at about 1500 N. Typically core indentation and initiation of delaminations occur at this load level. The peak core shear stress at the rear face sheet interface does however not change its location while its magnitude increases with rising thermal loads. Figure 6.5 shows the maximum core shear stress distributions of impacts without a thermal load and with the maximum thermal load of -225 K equivalent to an operating temperature of -55 °C.

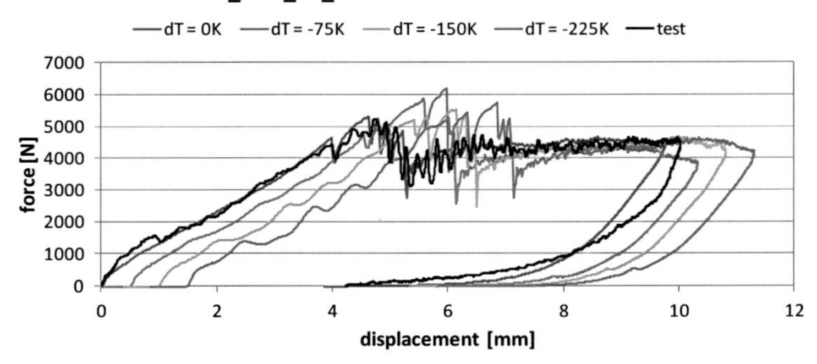

Figure 6.3 – impact response of test and simulations with different thermal loads

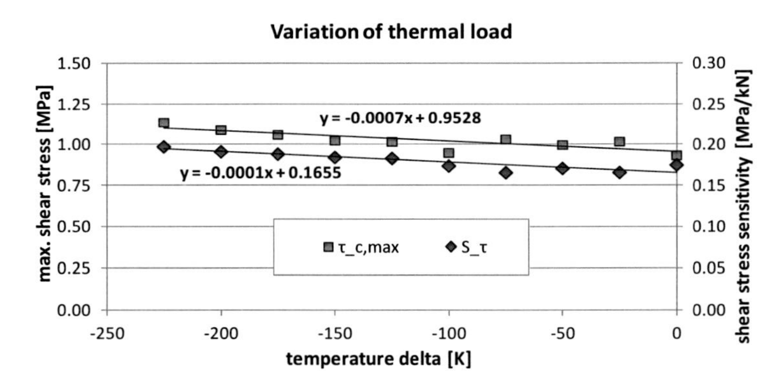

Figure 6.4 – shear stress sensitivity: Variation of thermal load

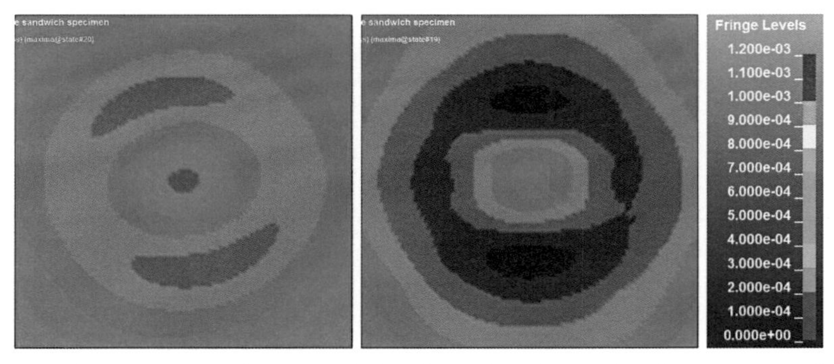

Figure 6.5 – maximum core shear stress in GPa at the rear face sheet interface during simulation of a 35 J impact on specimen RT_25.7_P2; thermal loads of 0 K (left) and -225 K (right) were applied; area shown is 100 x 100 mm with a centrally placed impact

6.3 Boundary conditions

The second investigated impact parameter were the boundary conditions (BC) of the sandwich plate. This investigation was split into two studies. Both studies were based on the same sandwich configuration RT_25.7 but varied in the first case specimen size and in the second case impact location. Both studies compared additionally the effect of clamped and simply supported BCs. The impact energy was fixed to 35 J sufficient for face sheet rupture. A thermal load of ΔT = -150 K was also applied. Figure 6.6 shows the different specimen sizes and investigated impact positions based on the reference specimen.

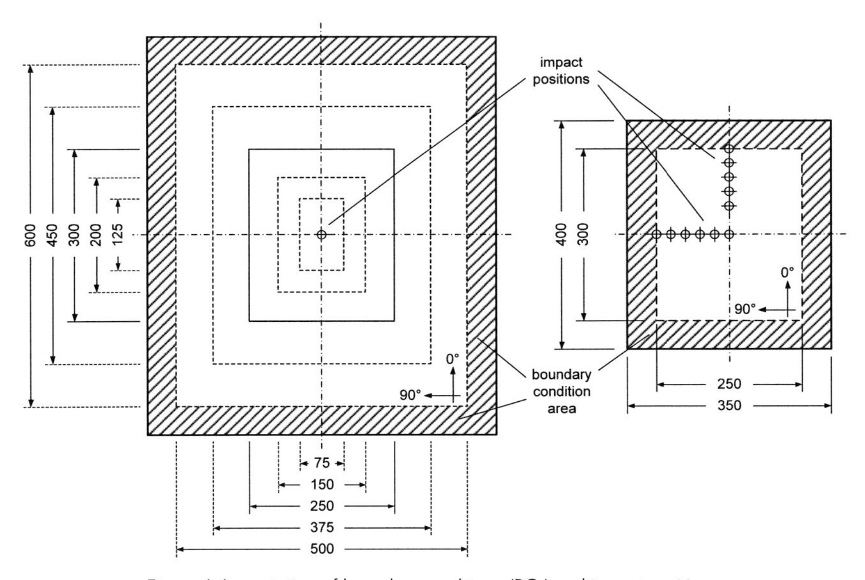

Figure 6.6 – variation of boundary conditions (BCs) and impact positions

Results of the first study are presented in Figure 6.7 using the absolute value of the maximum shear force $\bar{\tau}_c$ during impact and the shear stress sensitivity S_τ. Simply supported and clamped boundary conditions are distinguished. The distance to the boundary condition is always measured in the shortest direction, which is in this study always the 90° laminate orientation. Impact location was always the center of the specimen. Relevant for distinguishing the different specimens is not total specimen size but the size of the unsupported window in the center, as it defines the relevant boundary conditions. Window size was varied from 75 x 125 mm, which is equivalent to standardized impact testing of monolithic laminates [Air05], up to 500 x 600 mm, twice the size of the impact window used during experimental testing in chapter 4. 150 x 200 mm, 250 x 300 mm and 375 x 450 mm were selected as intermediate window sizes as shown in Figure 6.7. Total specimen size of the sandwich plate was always selected at least 50 mm larger to avoid edge effects or influences from the open sandwich run out due to thermal stresses.

The results show that very small sandwich specimens of the size used for impact testing of monolithic laminates are subject to significantly higher core shear stresses than larger sandwich specimens. Similarly the shear stress sensitivity S_τ drops from a maximum of 0.245 MPa/kN and 0.240 MPa/kN for the smallest specimen size with simply supported and respectively clamped BCs to 0.180 MPa/kN and 0.176 MPa/kN for the largest specimen. It is noted that the shear stress sensitivity is almost identical for the three largest specimens thus indicating that this value is a representative far field value. Thus the impact test setup and specimen size used during experimental testing in chapter 4 provide a good compromise between accuracy and specimen size.

In contrast the investigated BCs do not affect the shear stress sensitivity severely. The maximum shear stress sensitivity of 0.245 MPa/kN was observed with the smallest simply supported specimens. This is only slightly higher than 0.240 MPa/kN for the clamped specimen. Increasing the window size to 150 x 200 mm led to a significant decay of S_τ to 0.184 MPa/kN and 0.181 MPa/kN for simply supported and respectively clamped BCs. For the reference specimen with a 250 x 300 mm window S_τ was 0.185 MPa/kN and 0.180 MPa/kN for simply supported and respectively clamped BCs.

Figure 6.8 shows the distribution of the shear stresses for the three smallest specimen sizes with both simply supported and clamped BCs. Here it becomes clear how the BCs affect the shear stress distribution of the smallest specimen. In case of the 75 x 125 mm window the BCs lead to a severe interaction with the shear load resulting in two clearly visible maxima. The influence of the BCs decays significantly for the next larger window of 150 x 200 mm while for the 250 x 300 mm window there is hardly any influence anymore noticeable. It is thus concluded that small specimens are more sensitive to impact generated shear loads than large specimens as BCs and the impact shear load may interact.

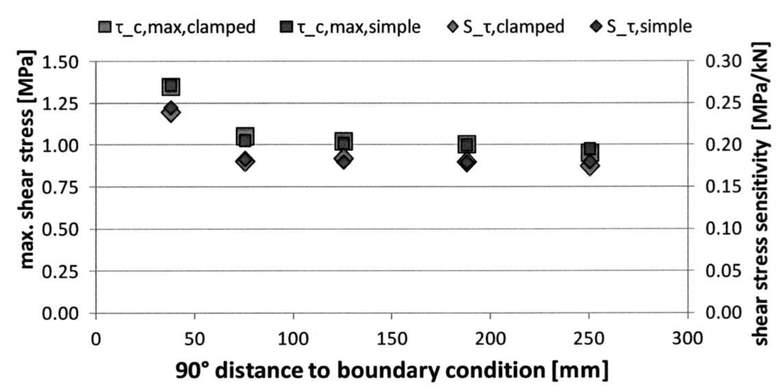

Figure 6.7 – shear stress sensitivity: Variation of specimen size with clamped and simply supported BCs; the shortest distance to the BC is measured in the 90° laminate orientation

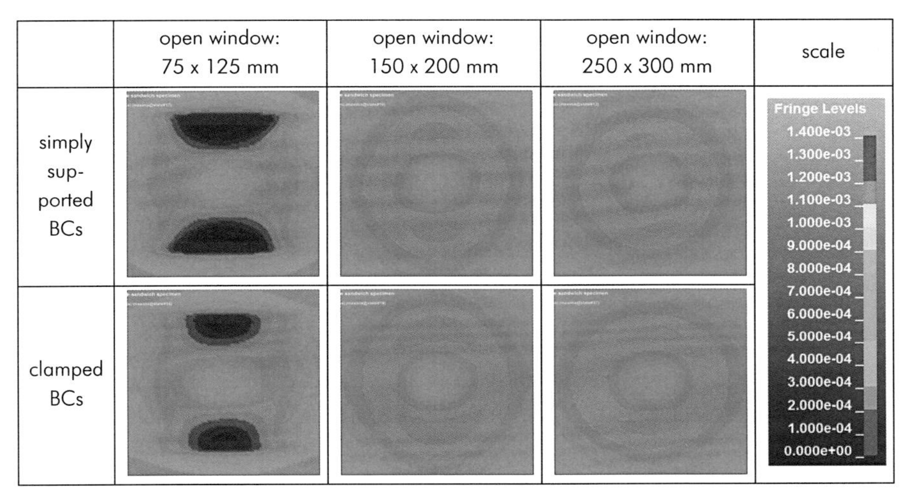

Figure 6.8 – maximum core shear stress in GPa: Variation of BCs and open window size; area shown is 100 x 100 mm with a centrally placed impact; upper row: simply supported BCs; lower row: clamped BCs; open window size: 75 x 125 mm (left), 150 x 200 mm (center) and 250 x 300 mm (right)

Referring to the shear stress sensitivity S_τ in Figure 6.7, the influence of the BC type is less clear and only of minor size. The smallest specimen is more sensitive to impact shear loads when embedded on simply supported BCs compared to clamped BCs. Increasing specimen size reverses this trend for intermediate sized sandwich specimens while larger specimens represent true far field values that appear to be hardly affected by the boundary conditions at all. The results of the maximum shear force $\overline{\tau}_c$ follow the same trends but show more scatter related to the variation of the maximum impact force.

The second study utilizes a fixed impact window size of 250 x 300 mm. Here the impact location was not anymore the center of the specimen but was instead varied in the 0° and 90° laminate directions in steps of 25 mm. Additionally both simply supported and clamped boundary conditions were investigated. Figure 6.9 and Figure 6.10 summarize the results using the absolute value of the maximum shear force $\overline{\tau}_c$ during impact and the shear stress sensitivity S_τ.

The results show a similar trend as those previously presented in Figure 6.7. Now referring to Figure 6.9 the maximum shear stress sensitivity is encountered with $S_\tau = 0.251$ MPa/kN for an impact located 50 mm away from the boundary condition in the 0° laminate orientation. This drops significantly for distances of 75 mm and 100 mm. Here a distance of 150 mm in 0° orientation and 125 mm in 90° orientation respectively represents a central impact on the sandwich plate. Impacts directly on the simply supported BC are less shear sensitive while an impact directly on the edge of clamped BCs could not be tested. Clamped BCs require fixation of the impacted top face sheet. Variation of the impact location in the 90° orientation does not change the discussed trends significantly as shown in Figure 6.10.

It is thus concluded that impacts in the vicinity of a boundary condition are more severe than those located further away. For a better clarification Figure 6.11 shows the distribution of the maximum core shear stresses for impacts placed at a distance of 25, 50 and 75 mm away from the constrained edge of the sandwich in the 0° laminate orientation. This shows clearly the influence of a boundary condition in the proximity of an impact as the highest shear stresses develop directly at or next to the constrained edge.

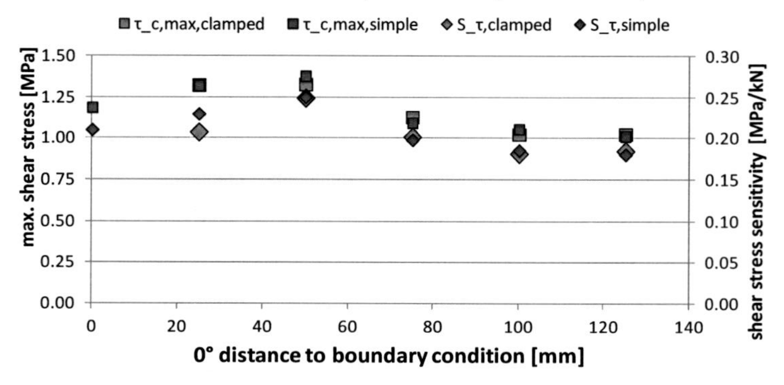

Figure 6.9 – shear stress sensitivity: Variation of impact location with clamped and simply supported BCs in 0° laminate orientation

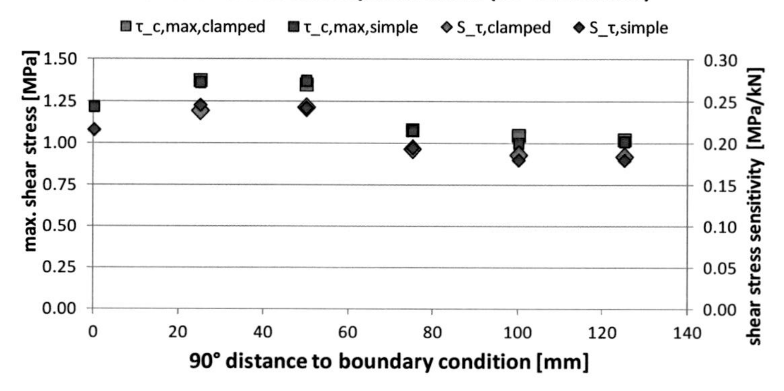

Figure 6.10 – shear stress sensitivity: Variation of impact location with clamped and simply supported BCs in 90° laminate orientation

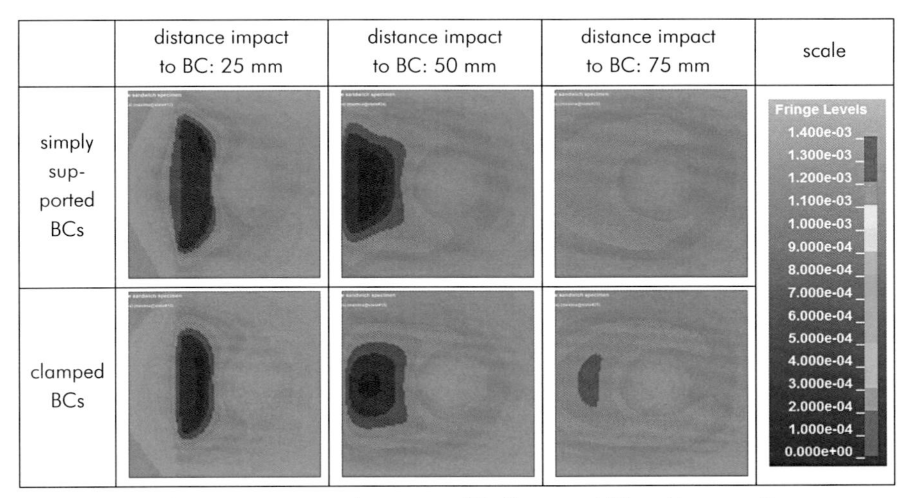

Figure 6.11 – maximum core shear stress in GPa: Variation of BCs and impact position; area shown is 100 x 100 mm with a centrally placed impact; upper row: simply supported BCs; lower row: clamped BCs; distance to BC is 75 mm (left), 50 mm (center) and 25 mm (right)

6.4 Impactor geometry and velocity

Impacts can be described by the characteristics of the impactor itself and by its dynamic properties. This can be further divided into contact properties such as the stiffness and shape of the impactor as well as the dynamic properties mass and velocity.

Contact properties typically distinguish between sharp and blunt impactors. The analysis methods introduced in chapter 3 and 5 however neglect the effect of impactor stiffness as part of the contact properties. Both the analytical failure criterion and the simulation model use an ideally stiff impactor. Blunt impacts are typically tested using a hemispherical steel impactor representing relevant impact threats such as stones, debris or dropped tools. The size of the impactor may however vary depending on the individual impact threat. In this work only impactors with 25.4 mm diameter were used. Testing norms for monolithic laminates use smaller impactors with 16 mm diameter [Air05] while Raju et al. [Raj08] test sandwich specimens with impactors of size 25.4 mm and 76.2 mm in diameter.

The effect of the impactor diameter on core shear failure was thus investigated as a measure of the degree of bluntness of the impactor. Using equation (3.42) the radius of the impactor r_{imp} was varied from 8 mm to 38.1 mm representing the previously discussed impactor sizes. The applied material properties are identical to those in chapter 3. The results are shown in a failure mode map in Figure 6.12 and highlight that a more blunt impactor increases the foam core thickness necessary to prevent core shear failure.

According to the face sheet rupture criterion in equation (3.38) a blunter impactor distributes the load on a greater area of the face sheet which in turn increases the face sheet rupture force. This increases now the core shear stress and thus requires a thicker core for prevention of core shear failure.

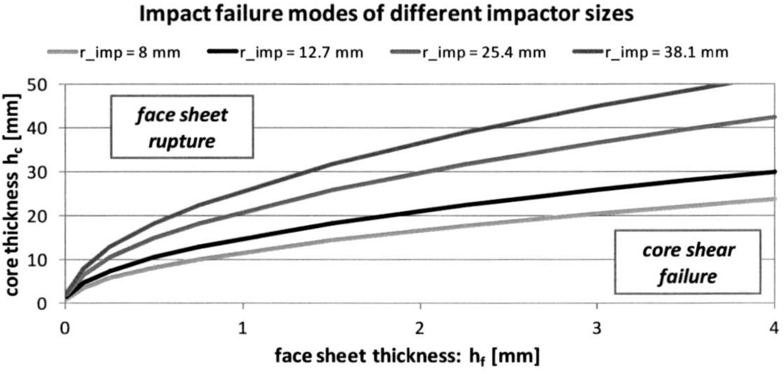

Figure 6.12 – impact failure mode map with different impactor sizes

The dynamic impact properties are categorized by mass and velocity of the impactor and often summarized by the single parameter impact energy. As described in section 3.3 the impact response of the sandwich varies depending on the ratio of the impacting mass relative to the mass of the impacted sandwich plate. The analytical failure criterion for core shear failure of Olsson and Block [Ols13] and the failure mode maps in section 3.5 are based on large mass and thus low velocity impacts, which neglect dynamic effects. These criteria therefore do not allow an investigation of the dynamic properties. Instead an analytical model based on the wave controlled impact response is required.

Testing in chapter 4 was based on large mass and thus quasi-static impact conditions using a reference impact energy of 35 J energy and a corresponding impact velocity of 4.77 m/s. The tests revealed an energy threshold required for each sandwich failure mode. If the impact energy is below this threshold, the sandwich will sustain only local damage, but not fail.

The simulation model developed in chapter 5 is based on static material characterization and validated against large mass impact tests from chapter 4. This model is generally capable to describe a wave controlled impact response as long as the mesh is sufficiently fine relative to the wave length. Additionally strain rate dependent material properties are included. Thus the simulation model can in principal describe dynamic effects. It must however be noted that the dynamic response was not validated for impact velocities beyond the available test results.

The simulation model was now used for a sensitivity study of the impact velocity by varying it for a 35 J impact from 2.38 m/s (half the reference velocity) to 23.85 m/s (five times the reference velocity). As the impact energy E_{kin} was kept constant, altering the impact velocity v_{imp} requires a change of the impactor mass m_{imp}. Table 6.1 tabulates the impact parameters and results. The results S_τ and $\bar{\tau}_c$ are also presented graphically in Figure 6.13. Additionally Figure 6.14 and Figure 6.15 display the force vs. displacement data and Figure 6.16 the max. core shear stress distribution at the rear face sheet interface.

The shear stress sensitivity S_τ increases from 0.175 MPa/kN for v_{imp} = 2.34 m/s by about 15% to S_τ = 0.200 MPa/kN for v_{imp} = 14.31 m/s but then drops to S_τ = 0.181 MPa/kN for v_{imp} = 23.85 m/s. This appears at first contradicting. Investigating the max. impactor force F_{max}, which initiates face sheet rupture, and the peak core shear stress $\bar{\tau}_c$ in more detail provides some explanation to this. Raising the impact velocity initially leads to a modest increase of the face sheet rupture load and thus effectively F_{max}. This in turn raises the shear load and thus $\bar{\tau}_c$ and has thus no effect on the shear stress sensitivity which is defined as the quotient of $\bar{\tau}_c$ and F_{max}.

Table 6.1 – simulation results of impacts with different velocities

E_{kin} [J]	v_{imp} [m/s]	m_{imp} [kg]	F_{max} [N]	$\bar{\tau}_c$ [MPa]	S_τ [MPa/kN]
35	2.38	12.358	5.473	0.958	0.175
35	4.77	3.075	5.553	1.025	0.185
35	9.54	0.769	5.465	1.010	0.185
35	14.31	0.342	5.750	1.148	0.200
35	19.08	0.1923	5.658	1.089	0.192
35	23.85	0.1231	5.988	1.082	0.181

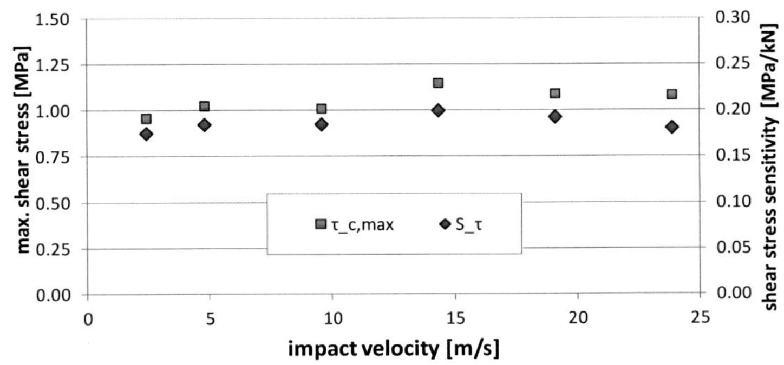

Figure 6.13 – shear stress sensitivity: Variation of impact velocity

This however interferes with dynamic effects due to wave propagation and reflection. Here attention should be given to Figure 6.16 which displays the distribution of the peak core shear stresses. For the lower velocities of $v_{imp} = 2.34$ m/s and 4.77 m/s, the peak core shear stresses describe a circle. Starting with $v_{imp} = 9.54$ m/s this circle shows distinct patterns of embedded smaller circles which change with each increase of the impact velocity and may not be explained by static load distribution alone. The peak shear stress sensitivity of $S_{\tau} = 0.200$ MPa/kN for $v_{imp} = 14.31$ m/s may thus be an effect of a wave interference phenomena either in the sandwich plane or through the thickness of the sandwich. This dynamic effect is also noticed in the impact response in Figure 6.14 and Figure 6.15. Here the displacement and load level, at which the first noticeable load drop occurs, increases with impact velocity. This is in contrast to the first interpretation of this load drop in chapter 4. Here this load drop was initially attributed to damage onset by core crushing and face sheet delaminations.

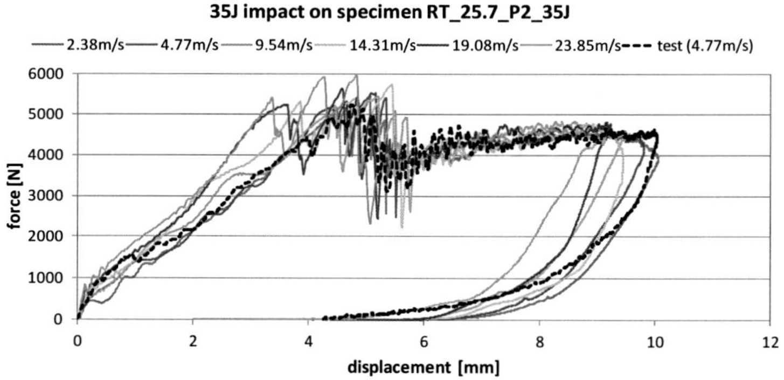

Figure 6.14 – sandwich impact response with different impact velocities

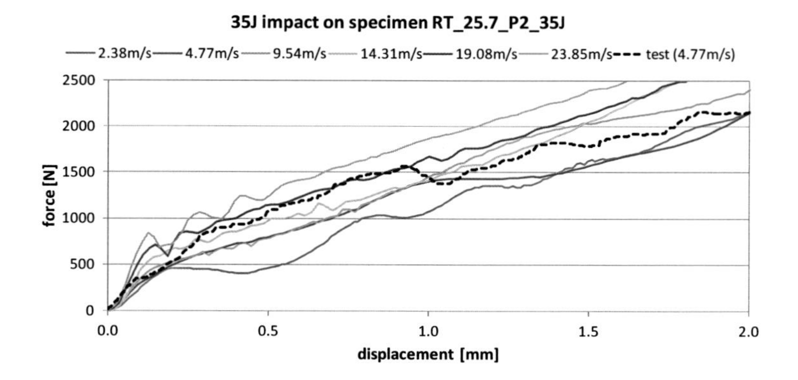

Figure 6.15 – initial sandwich impact response with different impact velocities

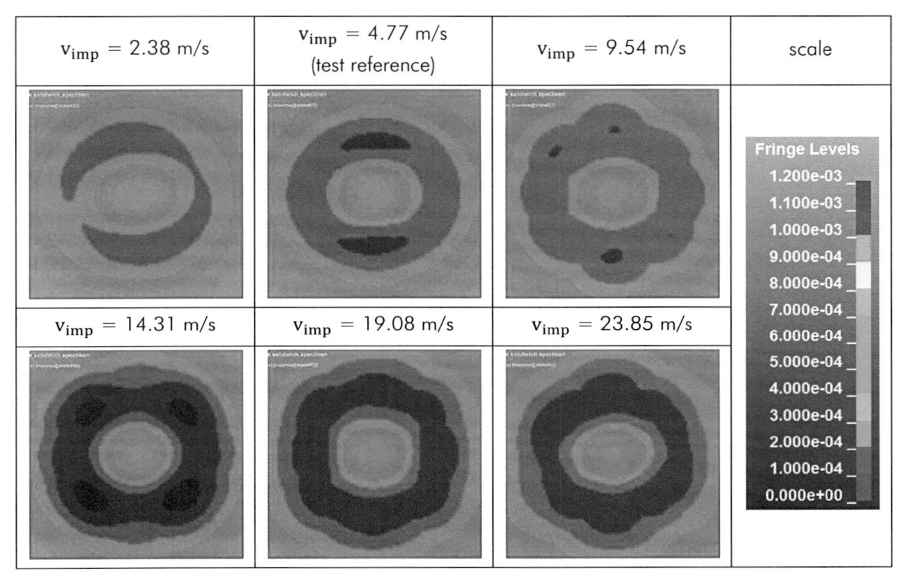

Figure 6.16 – maximum core shear stress in GPa: Variation of impact velocity; area shown is 100 x 100 mm with a centrally placed impact

6.5 Material properties of face sheet and core

From the perspective of a designer, the effect of the selected face sheet and core materials on the impact response are of particular interest to evaluate the benefit of using different materials for a particular component or structure. Also the thickness of a sandwich structure can be limited by the available design space. Thus different solutions are required for improvement of the sandwich impact resistance. The minimum core thickness required for prevention of core shear failure will be used in this context as a measure of the structure's sensitivity to core shear failure.

The face sheet properties stiffness summarized by the parameters E_r^* and v_r^* as well maximum fiber strain ε_{1T} influence the impact response. These parameters determine the face sheet rupture load F_{rup} as visible from equations (3.38) and (3.42). Generally speaking, a stronger face sheet increases the impact generated shear load of the foam core because the face sheet resists rupture longer compared to a weaker face sheet. If the normalized face sheet stiffness E_r^* is raised while maintaining the same failure strain ε_{1T}, the core shear load will increase. The same result will be obtained if the failure strain ε_{1T} increases and the face sheet stiffness E_r^* stays constant.

Different fiber types were compared to estimate the effect of the fiber material on the failure mode. The material properties of the carbon fibers were obtained from the manufacturer's data sheets [Ten12]. In case of E-glass material strength and failure strain vary significantly depending on the individual manufacturer or literature source [Zen95, Nie97, Sch07]. Thus a value of 3.5% was selected for ε_{1T} which corresponds to a moderate fiber strength of 2550 MPa. Laminate properties were obtained using the previously applied rules of mixture [Ikv04] without changing the matrix properties and the knockdown factor applied for NCF materials. The resulting material properties are summarized in Table 6.2 while Figure 6.17 shows the corresponding failure mode map.

Referring to Figure 6.17 the stronger UTS and IMS carbon fibers increase the face sheet rupture load as one would expect and thus require a thicker core to avoid core shear failure. The more brittle low cost STS carbon fiber is less critical due to its lower failure strain. The softer E-glass fiber is more shear critical due to its high failure strain of $\varepsilon_{1T} = 3.5\%$ and the related higher face sheet rupture load.

The designer may also change the matrix material. As the reinforcement fibers dominate laminate stiffness and strength, the effect of the matrix on the face sheet rupture force is limited to initiation and growth of delaminations, matrix cracking and interaction of these failure modes with fiber failure leading to face sheet rupture. The applied maximum strain failure criterion for face sheet rupture effectively neglects this. For a more detailed analysis an experimental investigation should be performed by e.g. a simplified approach based on quasi-static indentation tests comparable to section 5.4.2. This allows characterization of the face sheet rupture force F_{rup} of an individual face sheet core combination and avoids more complex impact tests. The effect of the matrix properties may then be analyzed by variation of the matrix material.

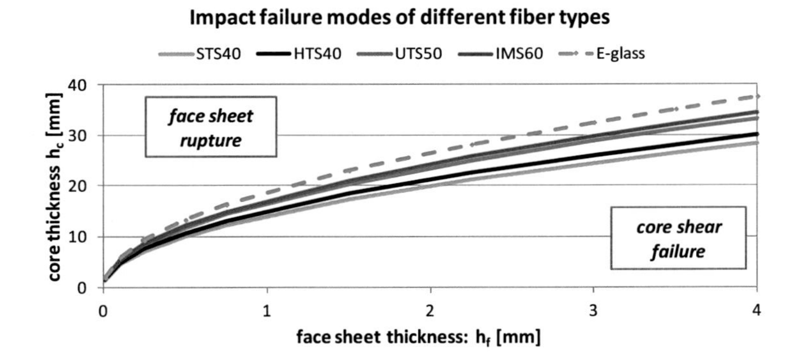

Figure 6.17 – impact failure mode map with different reinforcement fibers

Table 6.2 – face sheet material properties and impact properties for application of the analytical core shear failure criterion

fiber	resin	E_r^* [MPa]	ν_r^* [-]	ε_{1T} [%]	$\hat{\sigma}_{c,C}$ [MPa]	$\hat{\tau}_c$ [MPa]	R_{imp} [mm]	$\bar{a}^2$ [-]
E-glass	RTM-6	20650	0.295	3.5	1.8	1.4	12.7	0.8
Tenax STS40	RTM-6	48850	0.307	1.7	1.8	1.4	12.7	0.8
Tenax HTS40	RTM-6	48850	0.307	1.8	1.8	1.4	12.7	0.8
Tenax UTS50	RTM-6	48850	0.307	2.0	1.8	1.4	12.7	0.8
Tenax IMS60	RTM-6	57350	0.314	1.9	1.8	1.4	12.7	0.8

The effect of the core properties on the sandwich failure mode is according to equation (3.42) limited to the foam shear strength $\hat{\tau}_c$ and compressive strength $\hat{\sigma}_{c,C}$. The position of both material parameters in equation (3.42) describes clearly that increasing the shear strength improves the sandwich resistance against core shear failure, while increasing the compressive strength has the reverse effect and decreases the sandwich resistance against core shear failure. The effect of the shear strength will likely weigh heavier because the shear strength contributes directly to the minimum core thickness, while the compressive strength only contributes by its square root.

Structural foams can typically be purchased in different grades made of the same base material. These grades differ by their gaseous expansion resulting in several core densities as shown exemplary for the Rohacell RIST foam in Table 6.3. An increase of the foam density thus leads to higher compressive and shear strengths. The effect of a heavier foam core on the impact failure mode may thus not be judged immediately.

Figure 6.18 shows the governing failure mode of sandwich structures with different densities of the Rohacell RIST foam. The foam core material properties of Table 6.3 were used in combination with a CFRP face sheet with Tenax HTS40 carbon fibers and Hexcel RTM-6 resin as used previously for the mode transition curves in Figure 6.17. From this it can be concluded that increasing the core density improves the resistance of the sandwich structure against core shear failure. The effect of increasing the shear strength outperforms that of the increased compressive strength of the higher density foams. Optimization of a sandwich structure against a specific impact threat may in this context be performed using equation (2.22) which provides a way to describe the foam core properties relative to their density.

Table 6.3 – Rohacell RIST foam core properties [Evo13]

	ρ [kg/m³]	E_c [MPa]	G_c [MPa]	$\hat{\sigma}_{c,T}$ [MPa]	$\hat{\sigma}_{c,C}$ [MPa]	$\hat{\tau}_c$ [MPa]	ε_{max} [%]
51RIST	52	75	24	1.6	0.8	0.8	3.0
71RIST	75	105	42	2.2	1.7	1.3	3.0
110RIST	110	180	70	3.7	3.6	2.4	3.0

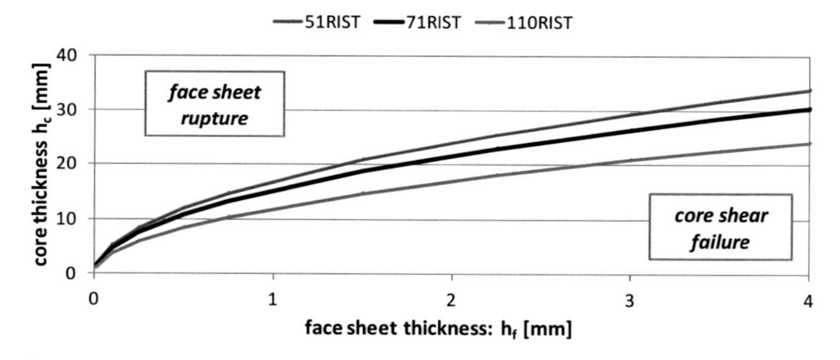

Figure 6.18 – impact failure mode map with different RIST foam core densities

An alternative way of enhancing the sandwich are foam core reinforcements. Endres [End10] developed the tied foam core (TFC) technology for textile reinforcement of foam cores and demonstrated it on Rohacell 31HF foam using monolithic CFRP pins. Published core properties are limited to the shear and compressive strength for a pin reinforcement angle of 45° relative to the sandwich plane. From this a best-fit line is provided. To make the TFC comparable to a pure foam core, strength properties were graphically red from the best-fit line at the reference densities 52, 75 and 110 kg/m³ and are summarized in Table 6.4 together with pure Rohacell HF foam data.

Figure 6.19 presents the resulting failure mode map. The low density Rohacell 31HF foam is very sensitive to shear failure due its low mechanical performance. The higher density foams 51HF and 71HF but also the reinforced foam cores are less sensitive similar as observed for the RIST foams. The minimum core thickness of the 31HF-TFC52 core is nearly identical to the equivalent density 51HF foam, while the heavier grade 31HF-TFC75 core slightly underperforms the 71HF foam. This simple comparison disregards however the greater strength of the TFC reinforced material which may lead to higher impact energy thresholds of the relevant failure mode. Furthermore it may be questioned whether a TFC reinforced foam core can be treated as a homogenous material, which a basic assumption of equation (3.42) used for creation of the failure mode map.

Table 6.4 – Rohacell HF foam and TFC reinforced HF properties [Evo13, End10]

	ρ [kg/m³]	E_c [MPa]	G_c [MPa]	$\hat{\sigma}_{c,T}$ [MPa]	$\hat{\sigma}_{c,C}$ [MPa]	$\hat{\tau}_c$ [MPa]	ε_{max} [%]
31HF	32	36	13	1.0	0.4	0.4	3.5
51HF	52	70	19	1.9	0.9	0.8	4.0
71HF	75	92	29	2.8	1.5	1.3	4.5
31HF-TFC52	52	–	–	–	2.4	1.3	–
31HF-TFC75	75	–	–	–	5.3	2.3	–
31HF-TFC110	110	–	–	–	9.8	3.8	–

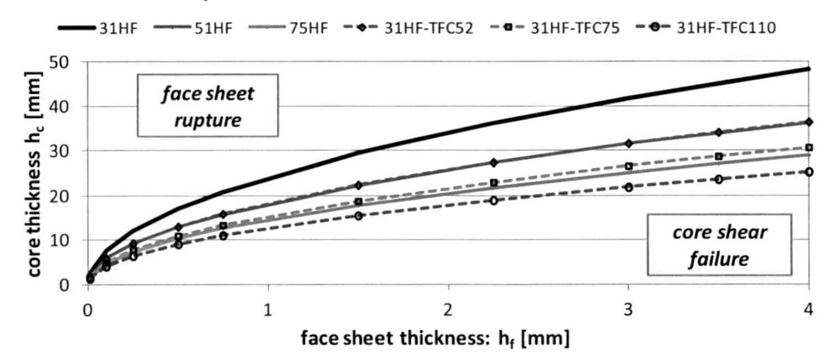

Figure 6.19 – impact failure mode map with pure and TFC reinforced foam cores

Another relevant material property is the interface fracture toughness expressed by the energy release rates G_{Ic} and G_{IIc}. The analytical model is purely strength based and thus determines the sandwich failure mode without consideration of the foam core fracture mechanics. The fracture toughness influences the numerical simulation predominantly by interface crack growth while crack initiation is primarily controlled by the foam core strength. A sensitivity study was thus performed with different fracture toughness values using a 20 J impact on the sandwich configuration RT_10.0. This is identical to specimen RT_10.0_P1 which is thus used as experimental reference. Figure 6.20 and Figure 6.21 show the impact response characterized by the force history and damage sizes for interface fracture toughness values of 100, 150 and 200 J/m².

The impact response of the 20 J impact is characterized by the initiation force of core shear failure and the following load drop. The fracture toughness influences the magnitude of this force despite the fact that the applied cohesive material model describes crack initiation based purely on material strength. Only crack growth is based on the fracture toughness. In fracture mechanics a minimum or critical crack size is required before critical and thus sudden crack growth occurs. Increasing the fracture toughness of a material increases the critical crack size, too [Gra03]. Applying this to the sandwich interface leads to the conclusion that the greater fracture toughness enlarges the allowable crack size and thus delays the load drop that is picked up by the impactor.

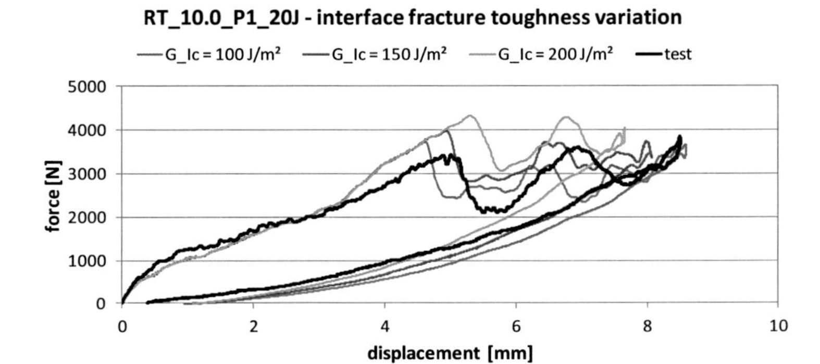

Figure 6.20 – impact response of test and simulations with different interface fracture toughness values

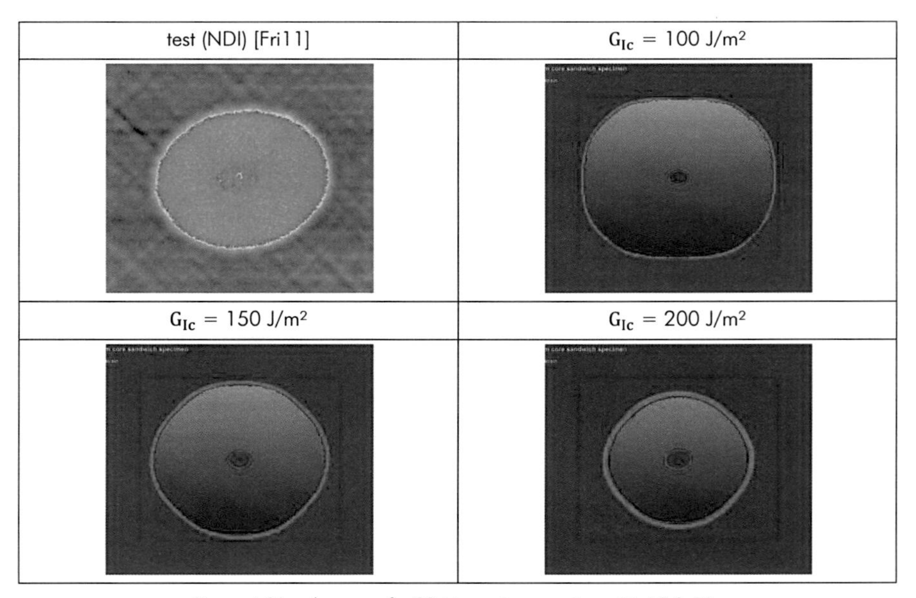

Figure 6.21 – damage of a 20 J impact on specimen RT_10.0_P1:
NDI image and simulated rear interface damage using different interface fracture toughness values;
color code – red, green (high attenuation, damage) and blue (low attenuation, no damage)

The simulation results in Figure 6.21 show that damage size decreases with increasing fracture toughness. Test and simulation agree best for an interface fracture toughness of 150 J/m^2. This contradicts the previous results of the impact response in Figure 6.20 where best agreement was reached for a value of 100 J/m^2. This contradiction may be explained by the simplified modeling approach that describes core shear failure as a rear face sheet debond. Crack initiation at the rear face sheet is driven in the simulation by the global core shear load while in the experiment stress concentrations at the indented upper face sheet may also contribute to crack initiation. Furthermore energy consumption due to crack growth through the core is not captured by simplified modeling approach.

6.6 Chapter summary

This chapter analyzed impact parameters with respect to their influence on the sandwich impact response. Focus was put on the impact failure mode utilizing the previously developed analytical failure mode map and the simulation model.

Fixing material and impact properties, the most important parameter that influences the sandwich impact response is the sandwich configuration. As expressed by equation (3.42), the critical and thus minimum core thickness $h_{c,min}$ that describes failure mode transition between core shear failure and face sheet rupture depends on the face sheet thickness h_f. As this relationship is nonlinear, a fixed ratio of the face sheet to core thicknesses cannot be given and instead a geometrical failure mode map is suited best to express the interaction.

Impact properties are divided into ballistic and contact properties. Contact properties are predominantly controlled by the impactor geometry. The analytical model is limited to round impactor geometries. Here a small diameter describes a sharp impactor while larger diameters describe more blunt impactors. Figure 6.12 shows that blunt impactors promote core shear failure and are thus considered more detrimental.

The ballistic properties include impact velocity and energy. For low velocity impacts with a quasi-static structural response, the ballistic properties are summarized in the single parameter impact energy. Impact energy has no immediate influence on the sandwich failure mode that will occur as it does not influence equation (3.42). However a minimum or threshold impact energy is required to trigger a particular failure mode. The response of impacts with less energy than the threshold will thus be largely elastic with only minor face sheet and core damage but no face sheet rupture or core shear failure.

The effect of the impact velocity on the sandwich failure mode was investigated with the numerical simulation model alone. Only minor influences were found up to about 25 m/s. It is however noticed that strain rate effects, which typically increase material strength and stiffness at the cost of ductility, lead to slightly higher face sheet rupture and core shear loads. In addition dynamic effects were observed to affect the core shear stresses of the reference sandwich configuration starting at about 10 m/s.

The boundary conditions of the sandwich plate have a diverse effect on the sandwich failure mode. Generally the global boundary conditions of the sandwich structure have a minor effect on the failure mode. Increasing the sandwich panel size leads to more elastic deformation and thus higher threshold energies but without a noticeable effect on the failure mode. This changes however significantly as the impact occurs in the vicinity of a constrained edge. Here the core shear load increases by up to 40%. In consequence sandwich impact is generally considered as a predominantly local problem with limited effects of the global panel behavior on the sandwich failure mode.

Thermal stresses in the sandwich emerge and can add to mechanical loads of the impact. This leads to an increase of about 12% of the critical core shear load at room temperature and 18% at temperatures of -55°C compared to a stress free sandwich.

Material properties of the face sheet and core also affect the sandwich failure mode. Of these the relevant core material properties are the compressive and shear strength. A higher compressive strength of the core may actually be detrimental. It increases the foam core thickness required to prevent core shear failure as it concentrates the load on a smaller area thus initiating shear failure and crack growth more easily. An increase of the core shear strength is beneficial and reduces the required core thickness.

Two ways of enhancing the core properties were analyzed. The first is to substitute the original foam core with a heavier foam. The second way is the use of core reinforcements such as monolithic CFRP pins. In both cases the core strength properties for compression and shear loads increase, leading to a similar weight based performance with respect to the occurrence of core shear failure. It must be noted that the total strength values of the pin reinforced core are significantly higher compared to pure foams of the same weight which will likely affect threshold impact energies and damage sizes of the reinforced cores. The foam core fracture toughness influences predominantly the extent of the core damage due to shear failure but also has an influence on the initiation of core shear failure.

The face sheet properties and its rupture load are controlled by the reinforcing fibers. Increasing fiber stiffness and strength, which is in this case described by the failure strain, promotes core shear failure. Face sheet fracture mechanics properties are however neglected as the applied analytical model is purely strength based.

It is thus summarized that the impact response of composite foam core sandwich structures is highly complex as different material and sandwich failure modes may occur. The governing failure modes are face sheet rupture are core shear failure. These are dominated by the interplay of the face sheet and core strength properties and other affecting factors such as impact and boundary conditions. Design of an impact resistant sandwich structure thus requires a good balance of face sheet and core properties that are tailored to the load requirements. Changing one parameter of the sandwich configuration consequently requires adapting the other parameters, too.

7 Concluding remarks

7.1 Summary of performed work

This work investigated the impact response of CFRP foam core sandwich structures with focus on the sandwich failure mode. The dominating and thus governing failure modes of the sandwich are face sheet rupture and foam core shear failure. Impact tests were performed on flat sandwich panels with different sandwich configurations at room and low temperature conditions. Analytical and numerical tools were discussed and validated against test results. A failure mode map of the sandwich impact response was created as a quick reference for designers. A numerical simulation model of the sandwich impact process was developed using the explicit finite element software LS-DYNA and validated against test results. Influencing parameters of the sandwich impact response were investigated by means of simulation and the analytical failure mode map.

Analytical failure criteria for sandwich beams and related failure mode maps are beneficial tools for the development of sandwich structures. A failure mode map for the impact response extends these by providing the designer information on the minimum core thickness required to prevent core shear failure due to impact. It was generally concluded that the sandwich geometric configuration of face sheet and core thickness is the most influencing parameter. Increasing the core thickness lowers the core shear stress while increasing the face sheet thickness enlarges the same and thus the likeliness of core shear failure.

Impact testing of flat sandwich panels at room and low temperature conditions highlighted the influence of the sandwich configuration on the failure mode. Specimens were made of CFPR face sheets and Rohacell 71RIST foam core. Test results agree well with the failure mode map. Damage sizes of specimens with face sheet rupture were observed significantly smaller with a maximum damage diameter of 50 mm in comparison to 250 mm for core shear failure. Residual dent depth – which correlates with damage visibility – behaved contrary to this. Face sheet rupture caused greater residual dent depth and thus better visibility of the damage compared to core shear failure. Impact energy was observed to have no measurable effect on the sandwich failure mode, however a minimum impact energy – also referred to as threshold energy – is in any case required for initiation of either failure mode. Observed damages were finally categorized into four different classes of which the first corresponds to minor local damage, the second to face sheet rupture, the third to foam core shear failure and the fourth to thermal failure of the foam core.

Low temperature impact tests at -55 °C triggered vertical foam cracks as a new damage type. This damage, which is also referred to as thermal damage, was observed only in conjunction with face sheet rupture and severe core crushing that lead to a residual indentation depth of 2.5 mm or greater. The lower temperature also triggered a higher sensitivity of the panel against core shear failure. Sectioning of test specimens revealed a more brittle appearance of the core shear failure.

Development of a simulation method for sandwich impact was performed using the building block approach and highlights the complexity of failure of composite sandwich structures and the physical processes involved. The developed simulation strategy separates the intra- and interlaminar failure processes in the CFRP face sheets and splits the foam core behavior into an elasto-plastic core and a fracture mechanics controlled interface. Two material models representative for simulation of the interlaminar behavior of fiber reinforced composites were discussed, applied and calibrated against experiments. The Rohacell RIST PMI foam shows a complex failure behavior including quasi-plastic crushing during compressive loads and rather brittle fracture for tension and shear loads. A simplified yield surface was proposed for the elasto-plastic core behavior and compared to an invariant failure criterion developed particularly for this material class. Core interface failure is fracture mechanics controlled and was validated against experimental results. Validation of the full sandwich model was performed using sandwich indentation and impact tests. The model provided a good prediction of the governing failure mode and a reasonable representation of the impact damage.

The sensitivity of impact parameters was investigated with respect to the sandwich failure mode by means of simulation and analytical failure mode maps. This showed that when ignoring the effect of thermally dependent material properties, thermal stresses influence the sandwich failure mode moderately. This is somewhat in contrast to -55 °C impact tests, where vertical foam cracks emerged as a new damage type. Local boundary conditions had a noticeably stronger effect. Simulations of impacts located close to a constrained edge exhibited a local maximum of the core shear load. This peak load declines quickly as the impact is moved away from the support. Global boundary conditions had only a limited effect on the sandwich failure mode unless the panel becomes so small that the impact and the global boundary conditions interact. Larger panels absorb more impact energy by elastic deformation before failure is triggered. Thus the threshold energy increases in larger panels without a change of the sandwich failure mode that will occur if this energy is superseded. The effect of the impact velocity on the sandwich failure mode was investigated with the numerical simulation model alone. Only minor influences were found up to about 25 m/s. Contact properties are defined by the material stiffnesses and impactor geometry. Small but sharp impactors lead to a more punctual loading resulting in face sheet rupture, while large and more blunt impactors distribute their load stronger across the face sheet and thus increase the core shear load.

Material properties were investigated to clarify the effect of changing face sheet and core materials. It was concluded that increasing the strength and thus rupture load of the face sheet material also increases the core shear load and thus requires a similar strengthening of the core either by increasing its thickness or selecting a stronger material in order to keep the same characteristic failure mode. The foam core fracture toughness has only a limited effect on the failure mode but affects damage size more severely.

This summary has demonstrated that a series of complex damage processes occur during impact on composite sandwich structures. These can be reduced to two dominating failure initiating loads of which the first is the face sheet rupture force and the second the peak core shear load at the outer edge of the core crushing zone. In a well designed sandwich structure these loads and their respective strengths are in good balance comparable to the design process of a sandwich beam subject to three-point bending (3PB). As face sheet rupture is generally preferred, a conservative design should aim for sufficient dimensioning of the core shear strength relative to the face sheet strength.

7.2 Limitations and implications

Limitations of this study concentrate on the material properties of composites made of multiaxial non-crimp fabrics, consideration of their failure behavior, thermally dependent material properties and fracture mechanics of the foam core.

In the simulation model the CFRP face sheets were described as a laminate of UD plies while the analytical failure criterion considered a quasi-isotropic laminate. Applied material models for the intra- and interlaminar behavior of the face sheets do not consider specific characteristics of multiaxial NCFs. Thus the NCF material is described as a laminated prepreg material with material stiffness and strength values obtained from the NCF.

As the equivalent of a single UD ply of multiaxial NCF composite cannot be tested separately, its material properties were estimated based on rules of mixtures and UD prepreg material and then calibrated against test results of the NCF composite. Validation of the stiffness properties was performed with 3PB tests which are generally sensitive to the material thickness. Test specimens were taken from surplus sandwich specimens and thus contained residual resin from the sandwich interface which resulted in greater specimen thicknesses. In-plane tensile testing of the laminate avoids the difficulties involved with the bending test and is thus recommended for future work. It is however not possible to test a single NCF UD layer on its own in order to determine the material properties directly.

Fracture mechanics properties of the CFRP face sheet were unknown and thus taken from literature. In case of the interlaminar properties the fracture toughness was available for the same material combination of fiber and resin, which could be validated by interlaminar shear tests. The fracture toughness of intralaminar failure is however more difficult to determine. Here fiber fracture dominates the response. The fracture toughness values of the fiber dominated failure modes were determined by reverse engineering of 3PB tests. Literature results of similar materials were used as a starting point. Experimental determination of the fracture toughness of the fiber dominated failure modes can be performed by compact tension and compression specimens while those of the matrix dominated failure modes are typically obtained from interlaminar tests. Application of this to multiaxial NCF composites will have to deal with the stronger textile character of the material and limitations on the laminate stacking sequence. Generally a more robust engineering process for the determination of multiaxial NCF material properties is needed.

The analytical failure criterion is completely strength based and does not consider effects of fracture toughness. The simulation model can apply fracture toughness and thus fracture mechanics effects in the face sheet and the sandwich interface but not in the foam core itself. The simulation thus describes core shear failure as a predominantly strength dominated failure mode which agrees with impact test results. This may however be not true if interaction of core shear failure and core crushing damage takes place as the brittle appearance of the failed low temperature test specimen may suggest.

Thermal strains and related residual stresses in the sandwich are neglected in the analytical model but taken into account in the simulation model. Due to a lack of knowledge on material properties at low temperatures, this thermal constraint was only considered by application of thermal strains relative to the baseline stress free temperature, but without consideration of thermally dependent material properties. Finally stress relaxation, which is a common phenomenon of polymeric materials, was ignored.

Implications of this work are found in three areas: Impact testing of sandwich structures, design of impact resistant composite sandwich structures and damage tolerance analysis of the same.

Impact testing of monolithic laminates is standardized by test specimens of size 150 x 100 mm placed on a 125 x 75 mm open window [Air05]. This specimen size is generally not advised for testing of composite sandwich structures. Local boundary conditions affect the load distribution during the impact test and thus the resulting damage size and type significantly. As a standardized test should generally determine the impact behavior free of local effects, specimen size must be selected large enough to describe the far field response which thus applies to the open window for impact testing of composite sandwich structures and the related specimen size. For the sandwich configurations used in this study a specimen size of 350 x 400 mm is recommended with clamped boundary conditions in a picture frame test setup with an unsupported window of size 250 x 300 mm. It is further noted that in practice impact testing is often performed with the specimen placed on a continuous rigid support. This type of test suppresses the failure mode core shear failure and is thus only capable of determining the indentation response of a sandwich structure.

Application of the findings of this work to the design of damage tolerant composite sandwich structures will make this process more complex as in addition to performance characteristics also the sandwich failure mode must be considered. Impact performance of composite materials including composite sandwich structures is typically determined by visibility and compression after impact testing and thus considers the effect of damage but not the failure mode. The failure modes face sheet rupture and core shear failure differ in appearance and resulting damage sizes. It is thus recommended to design sandwich structures with face sheet rupture as impact failure mode due to the smaller damage sizes in combination with greater visibility. The designer may use the developed failure mode map for determination of the minimum core thickness. Additionally a safety factor should be added to account for thermal stresses and stress concentrations due to local boundary conditions.

For CFPR face sheets and a Rohacell RIST PMI foam core, the critical core shear stresses may increase by up to 40% due to local boundary conditions and 18% due to thermal stresses.

Design of damage tolerant structures requires consideration of undetected and thus unknown defects. Impacts are a common source of damage and the resulting damage size just below the visibility threshold is used as a starting value for interface crack growth analysis. Here the simulation model can be used to investigate different impact scenarios in a virtual test environment supporting the development of damage tolerant structures by provision of initial damage sizes.

7.3 Future work

Impact resistance is from an overall industrial perspective part of the larger design requirement damage tolerance. Hence impact damage must be considered in light of their effect on the structural performance which is typically performed by testing the compression after impact strength of damaged laminates or sandwich structures. Cycling loading may lead to crack growth in the sandwich interface as a particularly sensitive spot of the structure. Additionally impact damage in sandwich structures may affect the stability behavior, too. Currently analysis and testing of the stability behavior is typically performed on undamaged structures which thus ignore this effect. Thus one major issue for future research is to develop practicable engineering solutions for impact simulation and treatment of the related damage tolerance requirements for aerospace applications. This can be seen as a general long term goal. It should be kept in mind that experience from other applications – here the marine industry is mentioned particularly – may be leveraged.

Deficits on material modeling became apparent during the development of the simulation model. Treatment of NCF composites and their failure behavior is an open issue. This work applied material models which were developed for laminate composites with good agreement between simulation and test results but required a significant effort for material calibration. Previous work has shown that there is a link between textile manufacturing parameters of NCF composites, their mechanical properties and failure behavior [Edg06]. A more robust engineering process to characterize failure and fracture mechanics of NCF composites should thus be addressed in future studies.

Another more apparent deficit are fracture mechanics based failure models for structural foams that include the typical crushing response of this material class. In this work the crushing response and failure of the foam core where modeled separately as failure was limited to the sandwich interface, while the crushing response was described with an elasto-plastic material model of the foam core. This approach showed good results but is not capable to describe cracks that propagate through the foam core. Kraatz [Kra07] developed an invariant based yield criterion for PMI foam cores which can be used as a failure criterion. For a full description of the mechanical response this must be extended by consideration of the crushing response and its fracture toughness into a single material model. Impact testing at low temperatures also revealed that thermal stresses in combination with core damage may initiate thermal cracks which should be addressed in future research.

The numerical sensitivity study of the parameter impact velocity showed that up to 25 m/s dynamic effects influence the sandwich impact response noticeably, but with only limited effect on the sandwich failure mode. As this was a very limited numerical study and dynamic effects were observed in the force history, a more thorough investigation of these effects should be performed by comparing true quasi-static indentation and impact tests with dynamic tests of the same.

This work investigated low velocity impacts as one type of impact. Depending on the structural application different impact scenarios must be considered. Some such as e.g. stone chipping, runway and engine debris or bird and hail strike have significantly higher velocities and energies leading to a different mechanical response. Consideration of strain rate effects will be of higher importance.

It was demonstrated that geometric boundary conditions lead to significant shear stress concentrations in the foam core. Thus for application of composite sandwich structures design solutions have to be developed that address critical areas such as load introduction points, sandwich run outs or transitions to monolithic laminates.

Improvements of the structural performance can be achieved by introduction of new materials. In case of sandwich structures the development of core reinforcement technologies such as the TFC technology as shown by Endres [End10] is promising. Monolithic core reinforcements open up additional degrees of freedom for the designer and may thus be leveraged to improve the impact and damage tolerance behavior of composite foam core sandwich structures. Analysis of their impact response is thus of high interest.

References

[Abr97] S. Abrate, 1997, *Localized impact on sandwich structures with laminated facings*, Applied Mechanics Review 50(2), 69-82.

[Abr98] S. Abrate, 1998, *Impact on composite structures*, Cambridge University Press, Cambridge, United Kingdom.

[Air05] Airbus S.A.S., 2005, *Determination of Compressive Strength after Impact*, Airbus Test Method, AITM 1-0010, Issue 3, Blagnac (Toulouse), France.

[Air13] A. Airoldi, G. Sala, P. Bettini, A. Baldi, 2013, *An efficient approach for modeling interlaminar damage in composite laminates with explicit finite element codes*, Reinforced Plastics and Composites 32(15), 1075-1091.

[All69] H.G. Allen, 1969, *Analysis and Design of Structural Sandwich Panels*, Pergamon Press, Oxford, United Kingdom.

[All95] O. Allix, P. Ladevéze, A. Corigliano, 1995, *Damage analysis of interlaminar fracture specimens*, Composite Structures 31, 61-74.

[Alt96] H. Altenbach, 1996, *Einführung in die Mechanik der Laminat- und Sandwichtragwerke (in German)*, Deutscher Verlag für Grundstoffindustrie, Germany.

[Asp04] L.E. Asp, S. Nilsson, S. Singh, 2004, *Effects of stitch pattern on the mechanical properties of Non-crimp fabric composites*, Proceedings of the 11th European Conference on Composite Materials (ECCM 11), Rhodes, Greece.

[Baa09] J. Baaran, 2009, *Visual Inspection of Composite Structures*, EASA-Research Project/2007/3 Final Report, Institute of Composite Structures and Adaptive Systems, DLR (German Aerospace Center), Braunschweig, Germany.

[Bat10] M. Battley, M. Burman, 2010, *Characterization of Ductile Core Materials*, Sandwich Structures and Materials 12, 237-252.

[Bar01] J.E. Barbero, P. Lonetti, 2001, *Damage model for composites defined in terms of available data*, Mechanics of Composite Materials and Structures 8(4), 299-315.

[Ber89] M.L. Bernard, P.A. Lagace, 1989, *Impact Resistance of Composite Sandwich Plates*, Reinforced Plastics and Composites 8(5), 432-445.

[Ber04] C. Berggreen, 2004, *Damage Tolerance of Debonded Sandwich Structures*, PhD Thesis, Technical University of Denmark, Lyngby, Denmark.

[Bie11] T. Bielefeld, 2011, *Personal communication*, Yacht Teccon Engineering, Bremen, Germany.

[Bin05] Q. Bing, C.T. Sun, 2005, *Modeling and testing strain rate-dependent compressive strength of carbon/epoxy composites*, Composite Science and Technology 65, 2481-2491.

[Blo07] T.B. Block, 2007, *Life prediction for large center cracked stiffened fuselage panels under constant amplitude loading*, Studienarbeit (research paper) Nr. 775, Institute for Aircraft Design and Lightweight Structures, Technical University of Braunschweig, Germany.

[Blo11] T.B. Block, C. Brauner, M.I. Zuardy, A.S. Herrmann, 2011, *Advanced numerical investigation of the impact behavior of CFRP foam core sandwich structures*, 3rd ECCOMAS Thematic Conference on the Mechanical Response of Composites (ECCOMAS Composites 2011), Hannover, Germany, 21-23 September 2011.

[Bra11] C. Brauner, T.B. Block, A.S. Herrmann, 2011, *Meso-level manufacturing simulation of sandwich structures to analyze viscoelastic-dependent residual stresses*, Composite Materials 46(7), 783-799.

[Cam03] P.P. Camanho, C.G. Davila, M.F. de Moura, 2003, *Numerical Simulation of Mixed-Mode Progressive in Composite Materials*, Composite Materials 37(16), 1415-1438.

[Cam06] P.P. Camanho, C.G. Davila, S.T. Pinho, L. Iannucci, P. Robinson, 2006, *Prediction of in situ strengths and matrix cracking in composites under transverse tension and in-plane shear*, Composites Part A 37, 165-176.

[Cam12] C.J. Cameron, A.M. Clark, M. Battley, 2012, *Influence of Testing Methods on the Measured Properties of Polymeric Foams*, 10th International Conference on Sandwich Structures (ICSS 10), Université de Nantes, Nantes, France, 27-29 August 2012.

[Cha87] F.K. Chang, K.Y. Chang, 1987, *A Progressive Damage Model for Laminated Composites Containing Stress Concentrations*, Composite Materials 21, 834-855.

[Cha11] G.B. Chai, S. Zhu, 2011, *A Review of low-velocity impact on sandwich structures*, Proceedings of the Institution of Mechanical Engineers, Part L: Materials Design and Applications 225, 207-230.

[Cou28] R. Courant, K. Friedrichs, H. Levy, 1928, *Über die partiellen Differenzengleichungen der mathematischen Physik*, Mathematische Annalen (in German) 100(1): 32-74.

[Cun97] R.G. Cuntze, R. Deska, B. Szelinski, R. Jeltsch-Fricker, S. Meckbach, D. Huybrechts, J. Kopp, L. Kroll, S. Gollwitzer, R. Rackwitz, 1997, *Neue Bruchkriterien und Festigkeitsnachweise für unidirektionalen Faserkunststoffverbund unter mehrachsiger Beanspruchung – Modellbildung und Experimente*, Fortschrittsberichte VDI 5 (506), VDI Verlag, Düsseldorf, Germany.

[Dan06] I.M. Daniel, O. Ishai, 2006, *Engineering Mechanics of Composite Materials*, Oxford University Press, New York, USA.

[Dan09] I.M. Daniel, E.E. Gdoutos, Y.D.S. Rajapakse (ed.), 2009, *Major Accomplishments in Composite Materials and Sandwich Structures*, An Anthology of ONR Sponsored Research, Springer, New York, USA.

[Dav03] C.G. Dávila, P.P. Camanho, 2003, *Failure criteria for FRP laminates in plane stress*, Technical Report NASA/TM-2003-212663, National Aeronautics and Space Administration, USA.

[Dav04] G. Davies, R. Olsson, 2004, *Impact on composite structures*, Royal Aeronautical Society, Aeronautical Journal 108(1089), 541–564.

[Dug60] D.S. Dugdale, 1960, *Yielding of steel sheets containing slits*, Journal of the Mechanics and Physics of Solids 8, 100-104.

[Dyn12] DYNAmore GmbH and LSTC Inc., 2012, *LS-DYNA Support site*, http://www.dynasupport.com/howtos/element/shell-formulations

[Edg04] F. Edgren, P.H. Bull, L.E. Asp, 2004, *Compressive failure of impacted NCF composite sandwich panels – Characterization of the failure process*, Composite Materials 38(6), 495-514.

[Edg05] F. Edgren, L.E. Asp, 2005, *Approximate Analytical Constitutive Model for Non-Crimp Fabric Composites*, Composites Part A 36, 173-181.

[Edg06] F. Edgren, 2006, *Physically Based Engineering Models for NCF Composites*, PhD Thesis, Kungliga Tekniska Högskolan, Stockholm, Schweden.

[Edg08] F. Edgren, C. Soutis, L. E. Asp, 2008, *Damage tolerance analysis of NCF composite sandwich panels*, Composite Science and Technology 68(13), 2635-2645.

[End10] G. Endres, 2010, *Innovative sandwich constructions for aircraft application*, 9th International Conference on Sandwich Structures (ICSS 9), California Institute of Technology, Pasadena, USA, 14-16 June 2010.

[Evo13] Evonik Industries AG, 2013, *Rohacell® home, products & services*, http://www.rohacell.com/product/rohacell/en/products-services/pages/default.aspx

[Fal01] M.L. Falk, A. Needlemann, J.R. Rice, 2001, *A critical evaluation of cohesive zone models of dynamic fracture*, Journal de Physique IV, Proceedings 5, 43-50.

[Fat01] M.S.H. Fatt, K.S. Park, 2001, *Dynamic models for low-velocity impact damage of composite sandwich panels – Part A: Deformation*, Composite Structures 52, 335-351.

[Fat01a] M.S.H. Fatt, K.S. Park, 2001, *Dynamic models for low-velocity impact damage of composite sandwich panels – Part B: Damage initiation*, Composite Structures 52, 353-364.

[Fat10] M.S.H. Fatt, S. Sirivolu, 2010, A wave propagation model *for the high velocity impact response of a composite sandwich panel*, Impact Engineering 37, 117-130.

[Fer10] P. Feraboli, F. Deleo, B. Wade, M. Rassaian, M. Higgins, A. Byar, M. Reggiani, A. Bonfatti, L. DeOto, A. Massini, 2010, *Predictive modeling of an energy-absorbing sandwich structural concept using the building block approach*, Composites Part A 41, 774-786.

[Fri11] B. Friederichs, 2011, *Personal communication*, DLR Institute of Composite Structures and Adaptive Systems, Braunschweig, Germany.

[Fro92] Y. Frostig, M. Baruch, O. Vilnay, I. Sheinman, 1992, *Higher-order theory for sandwich beam behavior with transversely flexible core*, Engineering Mechachnics 118(5), 1026-1043.

[Ger05] S. Gerlach, M. Fiolka, A. Matzenmiller, 2005, *Modelling and analysis of adhesively bonded joints with interface elements for crash analysis (article in German)*, 4. LS-DYNA Anwenderforum, Bamberg, Germany.

[Ger08] R. Gerlach, C. R. Siviour, N. Petrinic, J. Wiegand, 2008, *Experimental characterization and constitutive modeling of RTM-6 resin under impact loading*, Polymer 49, 2728-2737.

[Gib82] L.J. Gibson, M.F. Ashby, 1982, *The Mechanics of Three-Dimensional Cellular Materials*, Proc. of the Royal Society of London, Series A 382 (1782), 43-59.

[Gib88] L.J. Gibson, M.F. Ashby, 1988, *Cellular Solids: Structure and Properties*, 2nd Edition, Cambridge University Press, Cambridge, UK.

[Gra03] A.F. Grandt, 2003, *Fundamentals of structural integrity: damage tolerant design and nondestructive evaluation*, 1st Edition, John Wiley & Sons, Hoboken, New Jersey, USA.

[Grz10] M. Grzeschik, K. Drechsler, 2010, *Experimental Studies on Folded Cores*, 9th International Conference on Sandwich Structures (ICSS 9), California Institute of Technology, Pasadena, USA, 14-16 June 2010.

[Gut09] M. Gutwinski, R. Schäuble, 2009, *Einfluss von Temperaturwechseln auf die Struktureigenschaften von CFK-Schaum-Sandwichstrukturen*, in: 17. Symposium Verbundwerkstoffe und Werkstoffverbunde (ed W. Krenkel), Wiley-VCH, Weinheim, Germany.

[Has80] Z. Hashin, 1980, *Failure Criteria for Unidirectional Fiber Composites*, Applied Mechanics 47 (2), 329-335.

[Har83] J. Harding, L.M. Welsh, 1983, *A tensile testing technique for fiber-reinforced composites at impact rates of strain*, Material Sciences 18, 1810-1826.

[Har09] D. Hartung, 2009, *Material Behavior of Composites by three dimensional load conditions* (published in German), Doctoral Thesis, Technische Universität Carolo-Wilhelmina zu Braunschweig, DLR Forschungsbericht 2009-12, Braunschweig, Germany.

[Har12] S. Hartmann, 2012, *Introduction to Composite Material Modelling with LS-DYNA*, Dynamore GmbH, Stuttgart, Germany.

[Hei07] S. Heimbs, P. Middendorf, S. Kilchert, A.F. Johnson, M. Maier, 2007, *Experimental and Numerical Analysis of Composite Folded Sandwich Core Structures Under Compression*, Applied Composite Materials 14(5-6), 363-377.

[Hei08] S. Heimbs, 2008, *Sandwichstrukturen mit Wabenkern: Experimentelle und numerische Analyse des Schädigungsverhaltens unter statischer und kurzzeitdynamischer Belastung (in German)*, PhD Thesis, Institut für Verbundwerkstoffe, Technical University of Kaiserslautern, Kaiserslautern, Germany.

[Hei10] S. Heimbs, P. Middendorf, M. Klaus, S. Kilchert, A.F. Johnson, 2010, *Sandwich structures with textile-reinforced composite foldcores under impact loads*, Composite Structures 92, 1485-1497.

[Her05] A.S. Herrmann, P.C. Zahlen, I. Zuardy, 2005, *Sandwich Structures Technology in Commercial Aviation – Present Applications and Future Trends*, in: O. T. Thomsen (ed.): *Sandwich Structures 7: Advancing with Sandwich Structures and Materials*, 13-26, Springer.

[Het46] M. Hétenyi, 1946, *Beams on elastic foundations*, The University of Michigan Press, Ann Arbor, Michigan, USA.

[Hex12] Hexcel Composites, 2012, *Product data RTM 6*, http://www.hexcelcomposites.com.

[Hil48] R. Hill, 1948, *A theory of the yielding and plastic flow of anisotropic materials*, Proceedings of the Royal Society London A 193 (1033), 281-297.

[Hil76] A. Hillerborg, M. Modéer, P.E. Petersson, 1976, *Analysis of crack formation and crack growth in concrete by means of fracture mechanics and finite elements*, Cement and Concrete Research 6, 773-782.

[Hin04] M. Hinton, P. Soden, A.S. Kaddour, 2004, *Failure Criteria in Fibre-Reinforced-Polymer Composites*, Elsevier Science Ltd, Oxford, UK, 2004.

[Hin07] M. Hinton, A.S. Kaddour, 2007, *The second world wide failure exercise: Benchmarking of failure criteria under triaxial stresses for fibre-reinforced polymer composites*, 16th International Conference on Composite Materials (ICCM-16), Kyoto, Japan, 8-13 July 2007.

[Hin11] M. Hinton, A.S. Kaddour, S. Li, P.A. Smith, P. Soden, 2011, *Failure Criteria in Fibre Reinforced Polymer Composites: Can any of the Predictive Theories be Trusted?*, NAFEMS World Congress, Boston, USA, 23-26 May 2011.

[Hir02] A. Hirth, P. Du Bois, K. Weimar, 2002, *A Material Model for Transversely Anisotropic Crushable Foam in LS-DYNA*, 12th International LS-DYNA Users Conference, Detroit, MI, U.S.A.

[Hof67] O. Hoffman, 1967, *The brittle strength of orthotropic materials*, Composite Materials 1, 200-206.

[Hos01] M.V. Hosur, J. Alexander, U.K. Vaidya, S. Jeelani, 2001, *High strain rate compression response of carbon/epoxy laminate composites*, Composite Structures 52, 405-417.

[Hsi98] H.M. Hsiao, I.M. Daniel, 1998, *Strain rate behavior of composite materials*, Composites Part B 29, 521-533.

[Ikv04] Institut für Kunststoffverarbeitung Aachen – IKV, 2004, *Compositor: Hilfsmittel zur Analyse von Laminaten aus Faserverbundwerkstoffen (in German)*; Version 2.1, Aachen, Germany.

[Joh09] M. John, M. Rinker, R. Schäuble, 2009, *Bruchmechanische Kennwertermittlung an Polymethacrylimid-Hartschäumen (in German)*, 12th Symposium „Problemseminar Deformation und Bruchverhalten von Kunststoffen", Merseburg, Germany.

[Joh11] M. John, R. Schlimper, M. Rinker, T. Wagner, A. Roth, R. Schäuble, 2011, *Long Term Durability of CFRP Foam Core Sandwich Structures*, CEAS Aeronautical Journal 2(1-4), 213-221.

[Kad07] A.S. Kaddour, M. Hinton, S. Li, P.A. Smith, 2007, *Damage theories for fibre-reinforced polymer composites: The third world wide failure exercise (WWFE-III)*, 16th International Conference on Composite Materials (ICCM-16), Kyoto, Japan, 8-13 July 2007.

[Kad13] A.S. Kaddour, M. Hinton, 2013, *Maturity of 3D failure criteria for fibre-reinforced composites: Comparison between theories and experiments: Part B of WWFE-II*, Composite Materials 47(6-7), 925-966.

[Kae08] L. Kärger, J. Baaran, J. Teßmer, 2008, *Efficient simulation of low-velocity impact on composite sandwich panels*, Computers and Structures 86, 988-996.

[Kla12] M. Klaus, H.G. Reimerdes, N.K. Gupta, 2012, *Experimental and numerical investigations of residual strength after impact of sandwich panels*, International Journal of Impact Engineering 44, 50-58.

[Koe11] H. Koerber, P.P. Camanho, 2011, *High strain rate characterization of unidirectional carbon-epoxy IM7-8552 in longitudinal compression*, Composites Part A 42, 462-470.

[Koe11a] H. Koerber, J. Xavier, P.P. Camanho, 2011, *High strain rate behavior of unidirectional carbon-epoxy IM7-8552 material characterization and modeling*, Proceedings of the 3rd ECCOMAS Thematic Conference on the Mechanical Response of Composites (ECCOMAS Composites 2011), Hannover, Germany, 21-23 September 2011.

[Koj05] S. Kojima, T. Yasuki, S. Mikutsu, T. Takatsudo, 2005, *A Study on Yielding Function of Aluminum Honeycomb*, 5th European LS-DYNA Users Conference, Birmingham, United Kingdom.

[Kno08] M. Knops, 2008, *Analysis of Failure in Fiber Polymer Laminates: The Theory of Alfred Puck*, Springer, Berlin, Germany.

[Kra07] A. Kraatz, 2007, *Anwendung der Invariantentheorie zur Berechnung des dreidimensionalen Versagens- und Kriechverhaltens von geschlossenzelligen Schaumstoffen unter Einbeziehung der Mikrostruktur (in German)*; PhD Thesis, Martin-Luther-Universität Halle-Wittenberge, Halle, Germany.

[Kro97] L. Kroll, W. Hufenbach, 1997, *Physically Based Failure Criterion for Dimensioning of Thick-Walled Laminates*, Applied Composite Materials 4, 321-332.

[Kru01] R. Krueger, T.K. O'Brien, 2001, *A shell/3D modeling technique for the analysis of delaminated composite laminates*, Composites Part A 32, 25-44.

[Kru04] R. Krueger, 2004, *VCCT Virtual crack closure technique: History, approach and applications*, Applied Mechanics Review 57(2), 109-143.

[Lac02] T.E. Lacy, I.K. Samarah, J.S. Tomblin, 2002, *Damage Resistance Characterization of Sandwich Composites using Response Surfaces*, FAA Report DOT/FAA/AR-01/71, Federal Aviation Administration, U.S. Department of Transportation, Washington D.C., USA.

[Lac02a] T.E. Lacy, I.K. Samarah, J.S. Tomblin, 2002, *Damage Tolerance Characterization of Sandwich Composites using Response Surfaces*, FAA Report DOT/FAA/AR-02/101, Federal Aviation Administration, U.S. Department of Transportation, Washington D.C., USA.

[Lad92] P. Ladevèze, E. Le Dantec, 1992, *Damage modeling of the elementary ply for laminated composites*, Composite Science and Technology 43, 257-267.

[Lei09] J. Leijten, H.E.N. Bersee, O.K. Bergsma, A. Beukers, 2009, *Experimental study of the low-velocity impact behavior of primary sandwich structures in sandwich structures in aircraft*, Composites Part A 40, 164-175.

[Li00] Q.M. Li, R.A.W. Mines, R.S. Birch, 2000, *The crush behavior of Rohacell-51WF structural foam*, International Journal of Solids and Structures 37, 6321-6341.

[Lib48] C. Libove, S.B. Batdorf, 1948, *A General Small-Deflection Theory for Flat Sandwich Plates*, NACA TN 1526, also in *NACA report 899*.

[Lim04] T.S. Lim, C.S. Lee, D.G. Lee, 2004, *Failure Modes of Foam Core Sandwich Beams under Static and Impact Loads*, Composite Materials 38(18), 1639-1662.

[Lst06] J.O. Hallquist (ed.), Livermore Software Technology Corporation – LSTC, 2006, *LS-DYNA Theory Manual*, Livermore, CA, USA.

[Lst13] Livermore Software Technology Corporation – LSTC, 2013, *LS-DYNA Keyword User's Manual*, Volume I, R.7, June 2013, Livermore, CA, USA.

[Lst13a] Livermore Software Technology Corporation – LSTC, 2013, *LS-DYNA Keyword User's Manual*, Volume II - Material Models, R.7, June 2013, Livermore, CA, USA.

[Mai07] P. Maimí, P.P. Camanho, J.A. Mayugo, C.G. Dávila, 2007, *A continuum damage model for composite laminates: Part I – Constitutive model*, Mechanics of Materials 39, 897-908.

[Mai07a] P. Maimí, P.P. Camanho, J.A. Mayugo, C.G. Dávila, 2007, *A continuum damage model for composite laminates: Part II – Computational implementation and validation*, Mechanics of Materials 39, 909-919.

[Mar05] A. Marasco, 2005, *Analysis and evaluation of mechanical performance of reinforced sandwich structures: X-Cor™ and K-Cor™*, PhD Thesis, Cranfield University, Cranfield, UK.

[Mat91] A. Matzenmiller, J. Schweitzerhof, 1991, *Crashworthiness Simulations of Composite Structures - A First Step with Explicit Time Integration*, in: Nonlinear Computational Mechanics: State of the Art (ed P. Wriggers), Springer, Berlin, Germany.

[Mat95] A. Matzenmiller, J. Lubliner, R.L. Taylor, 1995, *A constitutive model for anisotropic damage in fiber-composites*, Mechanics of Materials 20, 125-152.

[McH01] J. McHugh, J. Döring, W. Stark, 2001, *Ultraschallcharakterisierung von vernetzenden Epoxid-Harzen (in German)*, DGZfP-Jahrestagung 2001, Berlin, Germany.

[Man13] A. Manes, A. Gilioli, C. Sbarufatti, M. Giglio, 2013, *Experimental and numerical investigation of low velocity impact on sandwich panels*, Composite Structures 99, 8-18.

[Men98] E. Menéndez Álvarez, 1998, *Characterization of impact damage in composite laminates*, FFA TN 1998-24, The Aeronautical Research Institute of Sweden, Bromma.

[Meo05] M. Meo, R. Vignjevic, G. Marengo, 2005, *The response of honeycomb sandwich panels under low-velocity impact loading*, International Journal of Mechanical Sciences 47, 1301-1325.

[Men12] V.G.K. Menta, R.R. Vuppalapati, K. Chandrashekhara, D. Pfitzinger, N. Phan, 2012, *Manufacturing and mechanical performance evaluation of resin-infused honeycomb composites*, Reinforced Plastics and Composites 31(6), 415-423.

[Met11] C. Metzner, 2011, *Project internal communication*, document 8 VTPNG 2495 C4M_HS_FVC Übersicht_V2 (in German), September 2010.

[Mil02] N.N., 2002, *MIL-HDBK-17, Composite Materials Handbook, chapter 4: Building Block Approach for composite structures*, vol. 3, Rev. F., ASTM International, West Conshohocken (PA), USA.

[Min94] R.A.W. Mines, C.M. Worrall, A.G. Gibson, 1994, *The static and impact behaviour of composite sandwich beams*, Composites 25(2), 95-110.

[Moe12] E. Möhle, P. Horst, S. Kreling, K. Dilger, 2012, *Behavior of asymmetric sandwich shells under axial compressive loading*, 15th European Conference on Composite Materials (ECCM15), Venice, Italy, 24-28 June 2012.

[Moo00] R.C. Moody, A. J. Vizzini, 2000, *Damage Tolerance of Composite Sandwich Structures*, FAA Report DOT/FAA/AR-99/91, Federal Aviation Administration, U.S. Department of Transportation, Washington D.C., USA, January 2000.

[New59] N.M. Newmark, 1959, *A method of computation for structural dynamics*, ASCE, Engineering Mechanics 85 (EM3), 67-94.

[Ngu05] M.Q. Nguyen, S.S. Jacombs, R.S. Thomson, D. Hachenberg, M.L. Scott, 2005, *Simulation of impact on sandwich structures*, Composite Structures 67, 217-227.

[Nie97] G. Niederstadt, 1997, *Ökonomischer und ökologischer Leichtbau mit faserverstärkten Polymeren (in German)*, expert-Verlag, Renningen-Malmsheim, Germany.

[Ols94] R. Olsson, 1994, *Simplified theory for contact indentation of sandwich panels*, M.Sc. Thesis, Massachusetts Institute of Technology, Cambridge, USA.

[Ols96] R. Olsson, H.L. McManus, 1996, *Improved Theory for Contact Indentation of Sandwich Panels*, AIAA Journal 34(6), 1238-1244

[Ols00] R. Olsson, 2000, *Mass criterion for wave controlled impact response of composite plates*, Composites Part A 31, 879-887.

[Ols01] R. Olsson, 2001, *Analytical prediction of large mass impact damage in composite laminates*, Composites Part A 32, 1207-1215.

[Ols02] R. Olsson, 2002, *Engineering Method for Prediction of Impact Response and Damage in Sandwich Panels*, Sandwich Structures and Materials 4, 3-28.

[Ols03] R. Olsson, 2003, *Closed form prediction of peak load and damage onset under small mass impacts*, Composite Structures 59, 341-349.

[Ols06] R. Olsson, 2006, *Delamination threshold load for dynamic impact on plates*, International Journal of Solids and Structures 43, 3124-3141.

[Ols10] R. Olsson, 2010, *Analytical model for delamination growth during small mass impacts on plates*, International Journal of Solids and Structures 47, 2884-2892.

[Ols13] R. Olsson, T.B. Block, 2013, *Criteria for skin rupture and core shear cracking during impact on sandwich panels*, submitted abstract to the 19th International Conference on Composite Materials (ICCM-19), Montreal, Canada, 28 July – 2 August 2013.

[Pet00] A. Petras, M.P.F. Sutcliffe, 2000, *Indentation failure analysis of sandwich beam*, Composite Structures 50, 311-318.

[Pin06] S.T. Pinho, L. Iannucci, P. Robinson, 2006, *Physically-based failure models and criteria for laminated fibre-reinforced composites with emphasis on fibre kinking: Part I: Development*, Composites Part A 37: 63-73.

[Pin06a] S.T. Pinho, L. Iannucci, P. Robinson, 2006, *Physically-based failure models and criteria for laminated fibre-reinforced composites with emphasis on fibre kinking: Part II: FE implementation*, Composites Part A 37, 766-777.

[Pin06b] S.T. Pinho, P. Robinson, L. Iannucci, 2006, *Fracture toughness of the tensile and compressive fibre failure modes in laminated composites*, Composites Science and Technology 66, 2069-2079.

[Pla66] F.J. Plantema, 1966, *Sandwich Construction*, John Wiley & Sons, New York, USA.

[Puc92] A. Puck, 1992, *Ein Bruchkriterium gibt die Richtung an (in German)*, Kunststoffe, Vol. 82(7): 607-610.

[Puc96] A. Puck, 1996, *Festigkeitsanalyse von Faser-Matrix-Laminaten: Modelle für die Praxis (in German)*, Carl Hanser Verlag, München, Germany.

[Puc98] A. Puck, H. Schürmann, 1998, *Failure Analysis of FRP Laminates by means of physically based phenomenological models*, Composite Science and Technology, Vol. 58: 1045-1067.

[Puc02] A. Puck, J. Kopp, M. Knops, 2002, *Guidelines for the determination of the parameters in Puck's action plane strength criterion*, Composite Science and Technology 62, 371-378.

[Puc02a] A. Puck, H. Schürmann, 2002, *Failure Analysis of FRP Laminates by means of physically based phenomenological models*, Composite Science and Technology 62, 1633-1662.

[Puc07] A. Puck, M. Mannigel, 2007, *Physically based non-linear stress–strain relations for the inter-fibre fracture analysis of FRP laminates*, Composite Science and Technology 67, 1955-1964.

[Rah11] M. Rahammer, 2011, *Evaluation of quasi-static indentation models for predicting delamination in composite honeycomb sandwich panels*, Diploma Thesis, Stuttgart University.

[Raj08] K.S. Raju, B.L. Smith, J.S. Tomblin, K.H. Liew, J. Guarddon, 2008, *Impact damage resistance and tolerance of honeycomb core sandwich panels*, Composite Materials 42(4), 385-412.

[Raj09] K.S. Raju, J. Tomblin, 2009, *Damage Tolerance Evaluation of Full-Scale Sandwich Composite Fuselage Panels*, FAA Report DOT/FAA/AR-09/9, Federal Aviation Administration, U.S. Department of Transportation, Washington D.C., USA.

[Rin08] M. Rinker, P.C. Zahlen, R. Schäuble, 2008, *Damage and failure progression of CFRP foam core sandwich structures*, Proceedings of the 8th International Conference on Sandwich Structures (ICSS-8), University of Porto, Porto, Portugal, 6-8 May 2008.

[Rin08a] M. Rinker, M. Gutwinski, R. Schäuble, 2008, *Experimental and theoretical investigation of thermal stress in CFRP foam core sandwich structures*, Proceedings of the 13th European Conference on Composite Materials (ECCM-13), Royal Institute of Technology, Stockholm, Sweden, 2-5 June 2008.

[Rin10] M. Rinker, 2010, *Personal communication*, Email, August 2010.

[Rin11] M. Rinker, 2011, *Bruchmechanische Bewertung der Schadenstoleranz von CFK-Schaum-Sandwichstrukturen (in German)*, Shaker Verlag, Aachen, Germany.

[Rin11a] M. Rinker, M. John, P.C. Zahlen, R. Schäuble, 2011, *Face sheet debonding in CFRP/PMI sandwich structures under quasi-static and fatigue loading considering residual thermal stress*, Engineering Fracture Mechanics 78, 2835-2847.

[Roe08] J. Rösler, H. Harders, M. Bäker, 2008, *Mechanisches Verhalten der Werkstoffe (in German)*, 3rd Edition, Vieweg + Teubner Verlag, Wiesbaden, Germany.

[Rot98] A. Rotem, 1998, *Prediction of laminate failure with the Rotem failure criterion*, Composite Science and Technology 58, 1083-1094.

[Rot08] M.A. Roth, A. Kraatz, 2008, *New 3-D failure criterion for sandwich foam core based on PMI*, Proceedings of the 8th International Conference on Sandwich Structures (ICSS-8), University of Porto, Porto, Portugal, 6-8 May 2008.

[Rot10] M.A. Roth, F. Goldmann, 2010, *Ganzheitliche Betrachtung von Schaum-Sandwich-Bauweisen in der Luftfahrt (in German)*, 59. Deutscher Luft- und Raumfahrtkongress 2010 (DLRK 2010), German Society for Aeronautics and Astronautics (DGLR), Hamburg, Germany, 31 August – 2 September 2010.

[Sae08] E. Saenz, A. Roth, F. Roseli, X. Liu, R. Thomson, 2008, *Mode I fracture toughness of PMI sandwich core materials*, Proceedings of the 8th International Conference on Sandwich Structures (ICSS-8), University of Porto, Porto, Portugal, 6-8 May 2008.

[Sal11] M. von Salzen, 2011, *Test Report FVW-11-051: Faserverstärkte Kunststoffe, Bestimmung der Biegeeigenschaften nach DIN EN ISO 14125: 1998 (in German)*, Faserinstitut Bremen e.V., Bremen, Germany.

[Sal12] M. von Salzen, 2012, *Test Report FVW-12-017: Kohlenstofffaserverstärkte Kunststoffe, Bestimmung der scheinbaren interlaminaren Scherfestigkeit nach DIN EN 2563 (in German)*, Faserinstitut Bremen e.V., Bremen, Germany.

[Sch05] P.M. Schubel, J.-J. Luo, I.M. Daniel, 2005, *Low velocity impact behavior of composite sandwich panels*, Composites Part A 36, 1389-1396.

[Sch07] H. Schürmann, 2007, *Konstruieren mit Faser-Kunststoff-Verbunden (in German)*, Springer, Heidelberg, Germany.

[Sch09] M. Schubert, 2009, *Vergleich eines Resin Transfer Moulding Epoxidharzsystems mit einem Prepregharzsystem*, in: 17. Symposium Verbundwerkstoffe und Werkstoffverbunde (ed W. Krenkel), Wiley-VCH, Weinheim, Germany.

[Shi03] A. Shipsha, S. Hallström, D. Zenkert, 2003, *Failure mechanisms and modeling of impact damage in sandwich beams – a 2D approach: Part I – experimental investigation*, Sandwich Structures and Materials 5: 7-31.

[Shi03a] A. Shipsha, S. Hallström, D. Zenkert, 2003, *Failure mechanisms and modeling of impact damage in sandwich beams – a 2D approach: Part II – analysis and modeling*, Sandwich Structures and Materials 5, 33-51.

[Sie97] R.L. Sierakowski, 1997, *Strain rate effects in composites*, Applied Mechanics Review 50(12), 741-761.

[Ste09] U. Stelzmann, M. Hörmann, 2009, *Einführung in LS-DYNA (in German)*, CADFEM GmbH, Grafing b. München, Germany.

[Str00] I.J. van Straalen, 2000, *Comprehensive Overview of Theories for Sandwich Panels;* Proceedings DOGMA Workshop on modeling of sandwich structures and adhesively bonded joints, September 1998, published by IDMEC Porto, 2000, http://www.dogma.org.uk/vtt/modelling/modellingindex.htm.

[Swa72] S.R. Swanson, 1972, *Limits of quasi-static solutions in impact of composite structures*, Composites Engineering 2(4), 261-267.

[Tan07] N. Taniguchi, T. Nishiwaki, H. Kawada, 2007, *Tensile strength of unidirectional CFRP laminate under high strain rate*, Advanced Composite Materials 16, 167-180.

[Ten12] Toho Tenax Europe GmbH, 2012, *Produktprogramm für Tenax® Filamentgarn (in German);* Delivery programme for Tenax® filament yarn, http://www.tohotenax-eu.com/produkte/tenax-kohlenstofffasern/filamentgarn.html

[Tho92] O.T. Thomsen, 1992, *Analysis of Local Bending Effects in Sandwich Panels Subjected to Concentrated Loads*, in: K. A. Olsson and D. Weissman-Berman (eds.): *Sandwich Construction 2*: Proceedings of the 2nd International Conference on Sandwich Structures, Gainesville, U.S.A., 9-12 March 1992.

[Tho95] O.T. Thomsen, 1995, *Theoretical and experimental investigation of local bending effects in sandwich plates*, Composite Structures 30, 85-101.

[Tho00] O.T. Thomsen, 2000, *Modeling of multi-layer sandwich type structures using a higher order plate formulation*, Sandwich Structures and Materials 2, 331-349.

[Tho05] O.T. Thomsen, E. Bozhevolnaya and A. Lyckegaard (eds.), 2005, *Sandwich Structures 7: Advancing with Sandwich Structures and Materials*, Proceedings of the 7th International Conference on Sandwich Structures, Aalborg University, Aalborg, Denmark, 29-31 August 2005, Springer.

[Thy11] ThyssenKrupp AG, 2011, *InCar, Steifigkeitsoptimierter Sandwichwerkstoff (in German)*, http://incar.thyssenkrupp.com/7_01_028_Sandwich.html?lang=de.

[Tim70] S.P. Timoshenko, S. Woinowsky-Krieger, 1970, *Theory of Plates and Shells*, 2nd Edition, McGraw-Hill, London, UK.

[Tom99] J. Tomblin, T. Lacy, B. Smith, S. Hooper, A. Vizzini, S. Lee, 1999, *Review of Damage Tolerance for Composite Sandwich Airframe Structures*, FAA Report DOT/FAA/AR-99/49, Federal Aviation Administration, U.S. Department of Transportation, Washington D.C., USA.

[Tom01] J. Tomblin, K.S. Raju, J. Liew, B.L. Smith, 2001, *Impact Damage Characterization and Damage Tolerance of Composite Sandwich Airframe Structures*, FAA Report DOT/FAA/AR-00/44, Federal Aviation Administration, U.S. Department of Transportation, Washington D.C., USA.

[Tom02] J. Tomblin, K.S. Raju, J. F. Acosta, B.L. Smith, N.A. Romine, 2002, *Impact Damage Characterization and Damage Tolerance of Composite Sandwich Airframe Structures – Phase II*, FAA Report DOT/FAA/AR-02/80, Federal Aviation Administration, U.S. Department of Transportation, Washington D.C., USA.

[Tom04] J. Tomblin, K.S. Raju, G. Arosteguy, 2004, *Damage Resistance and Tolerance of Composite Sandwich Panels – Scaling Effects*, FAA Report DOT/FAA/AR-03/75, Federal Aviation Administration, U.S. Department of Transportation, Washington D.C., USA.

[Tom05] J. Tomblin, K.S. Raju, T. Walker, J.F. Acosta, 2005, *Damage Tolerance of Composite Sandwich Airframe Structures – Additional Results*, FAA Report DOT/FAA/AR-05/33, Federal Aviation Administration, U.S. Department of Transportation, Washington D.C., USA.

[Tor12] Toray Carbon Fibers America Inc., 2012, *Product information, Standard Modulus Carbon Fibers;* Technical Data Sheet No. CFA-001, T300, http://www.toraycfa.com/standardmodulus.html

[Tri87] T.C. Triantafillou, L.J. Gibson, 1987, *Failure Mode Maps for Foam Core Sandwich Beams*, Materials Science and Engineering 95, 37-53.

[Tsa72] W. Tsai, E.M. Wu, 1972, *A General Theory of Strength for Anisotropic Materials*, Composite Materials 5, 58-80.

[Tur07] A. Turon, C.G. Davila, P.P. Camanho, J. Costa, 2007, *An engineering solution for mesh size effects in the simulation of delamination using cohesive zone models*, Engineering Fracture Mechanics 74, 1665-1682.

[Ver11] J. Verhaeghe, J. Kustermans, 2011, *3D sandwich panel manufacturing with great properties as result*, JEC Composites Magazine 69, 45-47.

[Ver12] M.C. de Verdiere, A.A. Skordos, A.C. Walton, M. May, 2012, *Influence of loading rate on the delamination response of untufted and tufted carbon epoxy non-crimp fabric composites / Mode II*, Engineering Fracture Mechanics 96, 1-10.

[Wan13] J. Wang, A.M. Waas, H. Wang, 2013, *Experimental and numerical study of the low-velocity impact behavior of primary sandwich structures in aircraft*, Composite Structures 96, 298-312.

[Whi87] J.M. Whitney, 1987, *Structural Analysis of Laminated Anisotropic Plates*, Technomic, Lancaster, PA, USA.

[Wie08] J. Wiegand, 2008, *Constitutive modelling of composites under impact loading*, PhD thesis, University of Oxford, Department of Engineering Science, U.K.

[Yam78] S.E. Yamada, C.T. Sun, 1978, *Analysis of Laminate Strength and Its Distribution*, Composite Materials 12, 275-284.

[Yan82] S.H. Yang, C.T. Sun, 1978, *Indentation law for composite and laminates*, Proceedings of the Composite Materials: Testing and Design (Sixth Conference), ASTM STP 787, pp. 425-449, Cairo, Egypt, 27-29 December 1982.

[Yok11] T. Yokozeki, T.B. Block, A.S. Herrmann, 2011, *Effects of residual thermal stresses on the energy release rates of sandwich beams for debond characterization*, Reinforced Plastics and Composites 30(8), 699-708.

[Zah08] P.C. Zahlen, M. Rinker, C. Heim, 2008, *Advanced Manufacturing of Large, Complex Foam Core Sandwich Panels*, Proceedings of the 8th International Conference on Sandwich Structures (ICSS-8), University of Porto, Porto, Portugal, 6-8 May 2008.

[Zen95] D. Zenkert (ed.), 1995, *An Introduction to Sandwich Construction*, EMAS Publishing Co. Ltd., UK.

[Zen97] D. Zenkert (ed.), 1997, *The Handbook of Sandwich Construction*, EMAS Publishing Co. Ltd., UK.

[Zho06] D.W. Zhou, W.J. Stronge, 2006, *Low Velocity impact denting of HSSA lightweight sandwich panels*, International Journal of Mechanical Sciences 48, 1031-1045.

[Zho06a] D.W. Zhou, W.J. Stronge, 2006, *Modal frequencies of circular sandwich panels*, Sandwich Structures and Materials 8, 343-357.

[Zho07] G. Zhou, M. Hill, 2007, *Investigation of parameters governing the damage and energy absorption characteristics of honeycomb sandwich panels*, Sandwich Structures and Materials 9(4), 309-342.

[Zho07a] Y. Zhou, Y. Wang, S. Jeelani, Y. Xia, 2007, *Experimental study of tensile behavior of carbon fiber and carbon fiber reinforced aluminum at different strain rates*, Applied Composite Materials 14, 17-31.

A Appendix: Sandwich theory

A.1 Sandwich beam theory

Sandwich beams – the simplest sandwich structure – behave in many ways similar to monolithic I-beams which are used in great extent in civil engineering but also in various other engineering disciplines. The sandwich specific construction of two thin but stiff face sheets, which enclose a thick but soft core, initiates a division of labor between the two components. This division of labor is much similar to that of the web and flange of the I-beam. Within the I-beam the flanges carry the dominating in-plane tensile and compressive loads as a result of global bending. In consequence the connecting web in the center of the beam is predominantly loaded in shear. Similarly the core inside a sandwich structure is, when subject to global bending, loaded in shear while the face sheets are loaded in tension or compression. The bending stiffness and strength of both I-beams and sandwich structures is due to their specific geometry and design high compared to flat monolithic constructions.

For investigation of the flexural rigidity of a sandwich structure one has to move one step back to the basic problem of a beam subject to a constant bending load. Due to the loading, the beam is deformed at a curvature κ_x (the inverse of the bending radius R_x) as shown in Figure A.1 [Zen95].

Figure A.1 – beam subject to a bending load [Zen95]

The strain at a distance z from the neutral axis then becomes:

$$\varepsilon_x = \kappa_x z\ . \tag{A.1}$$

This means that the strain varies linearly with respect to z. The bending M_x moment within the beam can now be expressed as the integral of the stress in the direction of the beam $\sigma_x(z)$ multiplied by the local coordinate z across the thickness of the beam z:

$$M_x = \int \sigma_x z dz\ . \tag{A.2}$$

Using the expression

$$\sigma_x = \varepsilon_x E = \kappa_x z E \tag{A.3}$$

equation (A.2) can be reorganized as follows:

$$M_x = \int \sigma_x z dz = \kappa_x \int E z^2 dz = \kappa_x EI,$$
$$\text{where} = \int E z^2 dz = D\ . \tag{A.4}$$

The expression EI is also known as the flexural rigidity and is the product of the elastic modulus E of the material and the area moment of inertia I. As however the elastic modulus varies within a sandwich beam this expression cannot be used. Instead EI will be entitled D representing a general beam cross-section.

Using equation (A.4) the strain ε_x can now be expressed as

$$\varepsilon_x = \frac{M_x}{D} z\ . \tag{A.5}$$

Thus the strain still varies linearly with z over the cross-section.

In order to quantify the curvature κ_x w.r.t. the terms of displacement, the assumption of small deflections has to be made.

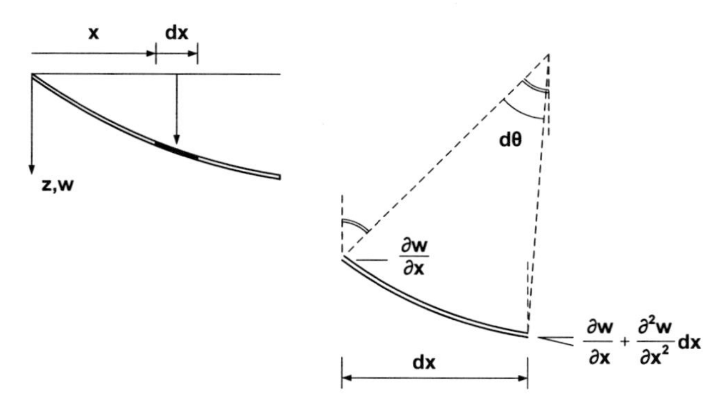

Figure A.2 – definition of the curvature κ_x [Zen95]

The geometrical study in Figure A.2 shows the following relationships [Zen95]:

$$dx = R\, d\theta, \tag{A.6}$$

$$d\theta = -\frac{d^2w}{dx^2}\, dx \tag{A.7}$$

Using those it is possible to define the curvature κ_x:

$$\kappa_x = \frac{1}{R_x} = -\frac{d^2w}{dx^2}\ . \tag{A.8}$$

Finally the cross-sectional properties of the sandwich beam have to be determined. These are summarized in the flexural rigidity D. Loads and boundary conditions as well as sandwich properties are now defined in Figure A.3 [Zen95].

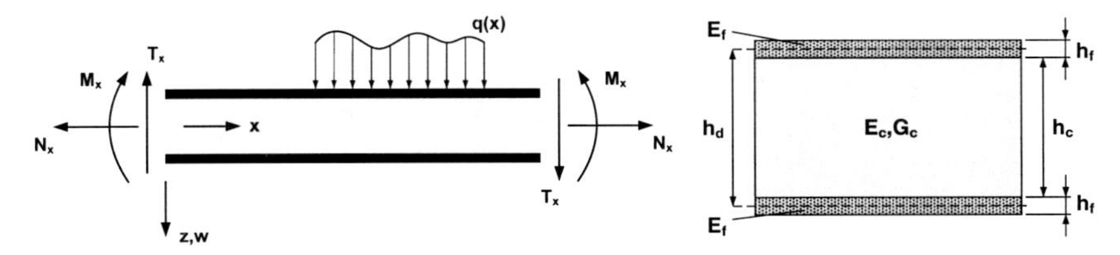

Figure A.3 – sign convention used for sandwich beams [Zen95]

Based on the selected definitions, the flexural rigidity D of the sandwich beam can now be expressed using the basic equation for a rectangular beam and the additional second moment of inertia of the face sheets as defined by the Huygens-Steiner theorem:

$$D = \int Ez^2 dz = \frac{E_f h_f^3}{6} + \frac{E_f h_f h_d^2}{2} + \frac{E_c h_c^3}{12} = 2D_f + D_0 + D_c \,. \tag{A.9}$$

Thus the flexural rigidity can now be separated into three constituents. The first D_f is the inertia of the face sheet itself, the second D_0 is the effect of placing the face sheets far away from the neutral fiber of the sandwich beam and the third D_c is the contribution of the core itself.

Typically sandwich structures are made of stiff engineering materials in the face sheets and rather light weight and thus weaker, more flexible core materials. Also the face sheets are usually much thinner than the core. Based on these observations the following assumptions can be made:

$$\text{Thin face sheet assumption: } h_f \ll h_c \,, \qquad \text{and the weak core assumption: } E_f \gg E_c \,. \tag{A.10}$$

Zenkert [Zen95] now shows, that with these two assumptions the margin of error will be less than 1% for each, if the following limits are maintained:

$$\text{Thin face sheet assumption: } \frac{h_d}{h_f} > 5.77 \,, \qquad \text{and the weak core assumption: } \frac{6E_f h_f h_d^2}{E_c h_c^3} > 100 \tag{A.11}$$

This simplifies equation (A.9) for the flexural rigidity of a sandwich beam significantly

$$D = D_0 = \frac{E_f h_f h_d^2}{2} \,. \tag{A.12}$$

Thinking about these assumptions and the resulting simplified equation (A.12) one major conclusion can be made. For typical sandwich structures with thin face sheets and a weak core more than 98% of the flexural rigidity is achieved solely due to the fact, that the face sheets are placed far away from the neutral axis of the sandwich beam and consequently loaded in tension and compression. This is often referred to as the sandwich effect, see also Figure 2.8. If however the connection of face sheet and core or the core itself fails, the sandwich structure loses its bending stiffness almost completely.

Based on Hook's material law describing linear elastic behavior and equation (A.5), which describes the strain distribution within the a sandwich beam, the normal stresses due to bending in the core and the face sheets can be described by

$$\sigma_f = \frac{M_x z E_f}{D} \quad \text{for } \frac{h_c}{2} \le z \le \frac{h_c}{2} + h_f \text{ , and}$$
$$\sigma_c = \frac{M_x z E_c}{D} \quad \text{for } z < \frac{h_c}{2} \text{ .} \tag{A.13}$$

Here σ_f is the normal stress in the face sheets due to bending and σ_c the same in the core. As pointed out by Zenkert [Zen95] the stresses vary linearly within each material but experience a jump at the interface between the two different materials. In-plane loading of the sandwich can be described simply by:

$$\varepsilon_{x0} = \frac{N_x}{E_{f1}h_{f1} + E_{f2}h_{f2} + E_c h_c} = \frac{N_x}{A_x} \text{ .} \tag{A.14}$$

Here ε_{x0} is the normal strain at the neutral axis. Thus the normal stresses can be determined by:

$$\sigma_f = \varepsilon_{x0} E_f \quad \text{and}$$
$$\sigma_c = \varepsilon_{x0} E_c \text{ .} \tag{A.15}$$

Stresses due to bending and in-plane loads of the sandwich beam can be superposed.

Similarly Zenkert describes the shear stresses in the sandwich. As they however do not vary linearly a more general formulation has to be found. For this an element dx of a sandwich beam as shown in Figure A.4 [Zen95] is considered.

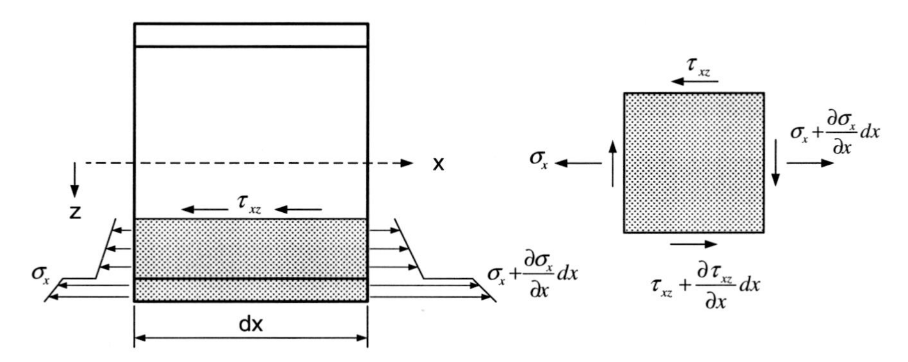

Figure A.4 – beam section dx defining equilibrium for a subsection [Zen95]

The shear force must now balance the change in the direct stress field:

$$\frac{\sigma_x}{d_x} + \frac{\tau_{xz}}{d_z} = 0 \quad \rightarrow \qquad \tau_{xz}(z) = \int_z^{(d+h_f)/2} \left(\frac{d\sigma_x}{dx}\right) dz \,. \tag{A.16}$$

By applying the first moment of inertia $B(z)$ Zenkert finally describes the shear stresses in the core and the face sheet with

$$\begin{aligned} \tau_c(z) &= \frac{T_y}{D}\left[\frac{E_f h_f h_d}{2} + \frac{E_c}{2}\left(\frac{{h_c}^2}{4} - z^2\right)\right] && \text{for } z < \frac{h_c}{2} \quad \text{and} \\ \tau_f(z) &= \frac{T_x}{D}\frac{E_f}{2}\left[\frac{{h_c}^2}{4} + h_c h_f + {h_f}^2 - z^2\right] && \text{for } \frac{h_c}{2} \le z \le \frac{h_c}{2} + h_f \,. \end{aligned} \tag{A.17}$$

Here $T_x = dM_x/dx$ and thus describes the change of the bending moment in the x-direction of the sandwich beam that is equivalent to a transverse force. Two important values for shear stresses can now be determined. The first is the maximum shear stress within the sandwich beam $\tau_{c,max}$. This is found in the center of the foam core. The second important value of interest is the shear stress at the interface between the face sheet and the core which is at the same time the minimum shear stress in the sandwich core $\tau_{c,min}$. By putting the respective coordinates of z into equation (A.17) these values can be determined as

$$\begin{aligned} &\tau_{c,max}(z=0) = \frac{T_x}{D}\left[\frac{E_f h_f h_d}{2} + \frac{E_c {h_c}^2}{8}\right] \qquad \text{and} \\ &\tau_{c,min} = \tau_{f,max} = \tau_c\left(z = \frac{h_c}{2}\right) = \frac{T_x}{D}\left[\frac{E_f h_f h_d}{2}\right] \,. \end{aligned} \tag{A.18}$$

The shear stress in the outer fiber of the face sheet is zero which can be seen from equation (A.17).

It can be concluded from equation (A.18) that the ratio between the minimum and maximum shear stress is less than one percent if

$$\frac{4E_f h_f h_d}{E_c {h_c}^2} > 100 \,. \tag{A.19}$$

As shown already in equations (A.10) there are typically two assumptions being made that simplify the analysis of sandwich structures. These can now be extended to the stress calculation, too. Assuming a weak core material $E_f \gg E_c$ the stresses can be rewritten as [Zen95]

$$\begin{aligned} &\sigma_c(z) = 0 \,, && \sigma_f(z) = \frac{M_x z E_f}{(D_0 + 2D_f)} \,, \\ &\tau_c(z) = \frac{T_x E_f h_f h_d}{2(D_0 + 2D_f)} \,, \text{ and} && \tau_f(z) = \frac{T_x}{2(D_0 + 2D_f)}\frac{E_f}{2}\left(\frac{{h_c}^2}{4} + h_c h_f + {h_f}^2 - z^2\right) . \end{aligned} \tag{A.20}$$

If additionally also the face sheets are thin and thus $h_f \ll h_c$, equations (A.20) simplifies further with [Zen95]

$$\sigma_c(z) = 0\,, \qquad \sigma_f(z) = \pm\frac{M_x}{h_f h_d}\,,$$
$$\tau_c(z) = \frac{T_x}{h_d}\,, \text{ and} \qquad \tau_f(z) = 0\,. \tag{A.21}$$

Based on these assumptions Zenkert draws a conclusion which helps to understand the principal division of work between the different constituents of a sandwich structure.

"This simplifies the modus operandi or the principal load carrying and stress distributions in a structural sandwich construction to: the faces carry bending moments as tensile and compressive stresses and the core carries transverse forces as shear stresses" ([Zen95] p.44).

The effect of the approximations as well as the discussed division of work between the different partners is also shown in Figure A.5 [Zen95].

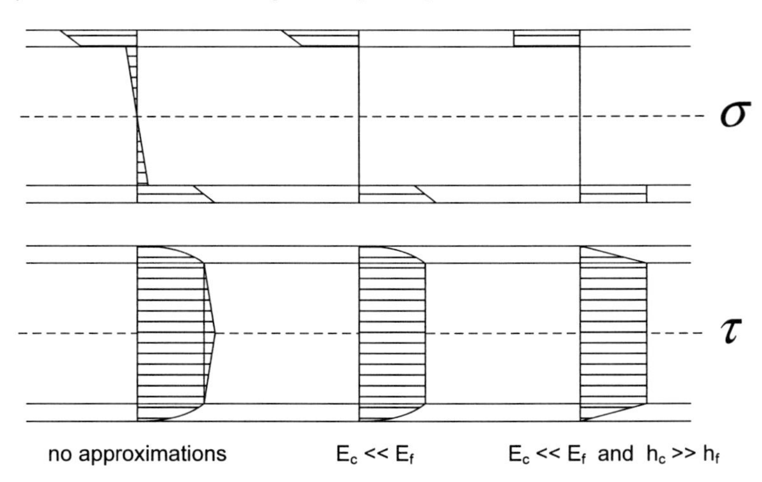

Figure A.5 – stresses due to global bending in the constituents of a sandwich beam representing the division of work [Zen95]

A.2 Sandwich plate theory

Sandwich plate theory can be defined using either the same notation as for sandwich beams and is thus similar to section A.2. or the notation typical from the classical lamination theory (CLT) used for the monolithic laminates. For the CLT the reader shall refer to common literature about the mechanics of laminated composite materials such as e.g. Niederstadt [Nie97], Daniel [Dan06] or Schürmann [Sch07]. In order to stay consistent with the previously used notation for sandwich beams, this will be used first to explain the necessary fundamentals. Afterwards the transition to the CLT based notation will be given as the later is applied for most sandwich impact models discussed in section 3. Figure A.6 now shows the sign convention selected by Zenkert [Zen95]. It is used without changes.

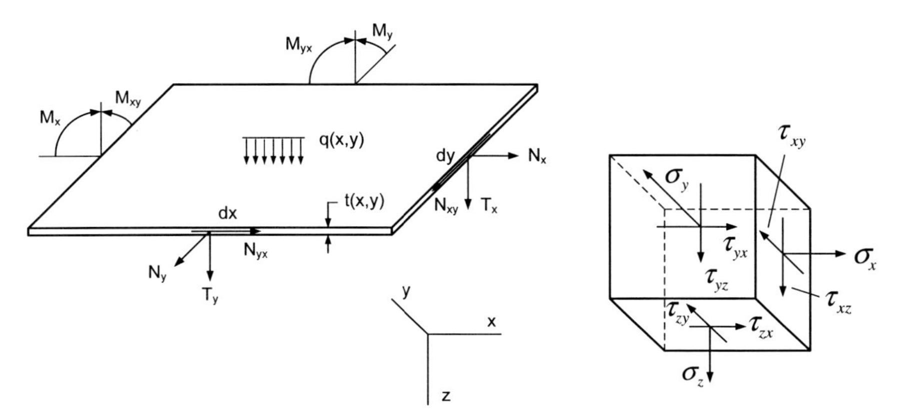

Figure A.6 – sign convention used for sandwich plate analysis [Zen95]

Zenkert [Zen95] describes a sandwich plate theory which is based on the basic small-deformation theory of Timoshenko and Woinowsky-Krieger [Tim70]. The later is extended in order to account for transverse shear deformation following work by Libove and Batdorf [Lib48]. Furthermore Zenkert introduces partial deflections similar to but simplified compared to the work of Plantema [Pla66]. Plantema separates the plate deflection w into their bending and shear deformation components but couples them in the plate x and y directions by

$$w = w_{bx} + w_{sx} = w_{by} + w_{sy} \ . \tag{A.22}$$

Zenkert simplifies this by discarding this coupling and instead simply superimposing the two displacements fields as done previously in the beam case. The deflection now becomes

$$w = w_b + w_s \ . \tag{A.23}$$

This simplification is valid only for fully isotropic sandwich structures as in orthotropic sandwich plates the bending and shear contributions differ depending on the plate orientation. The failure made for moderately orthotropic sandwich plates is however moderate according to Zenkert which is the reason the simplification is applied.

Skipping the derivation of the governing equations the cross-sectional properties of an orthotropic sandwich plate can be defined by [Zen95]

$$\begin{aligned} D_x &= \int z_x^2 \, E_x dz_x \approx \frac{E_{x1}h_1 \, E_{x2}h_2 \, h_d^2}{E_{x1}h_1 + E_{x2}h_2} \quad \text{and} \\ D_y &= \int z_y^2 \, E_y dz_y \approx \frac{E_{y1}h_1 \, E_{y2}h_2 \, h_d^2}{E_{y1}h_1 + E_{y2}h_2} \ . \end{aligned} \tag{A.24}$$

Figure A.7 shows the definition of the sandwich plate properties separately for the x and y directions.

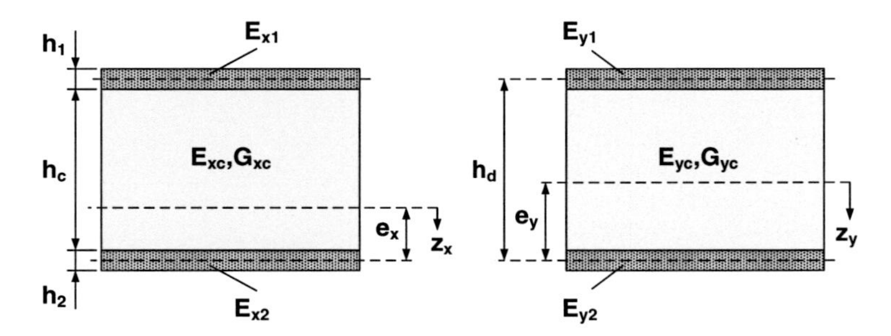

Figure A.7 – definitions and sign convention used for sandwich plate cross-section properties [Zen95]

The position of the neutral axis can differ in the x and y directions and is defined by

$$e_x = \frac{E_{x1} h_1 h_d^2}{E_{x1} h_1 + E_{x2} h_2} \quad \text{and} \quad e_y = \frac{E_{y1} h_1 h_d^2}{E_{y1} h_1 + E_{y2} h_2} . \tag{A.25}$$

Using equation (A.25) the neutral axis becomes identical in both the x and y direction when

$$\frac{E_{x2}}{E_{x1}} = \frac{E_{y2}}{E_{y1}} \quad \text{and thus} \quad e = e_x = e_y . \tag{A.26}$$

Assuming identical orthotropic laminate layups in both face sheets, the neutral axis wanders into mid-plane of the sandwich plate. This in combination with an isotropic core material such as foam leads to the following simplifications:

$$\begin{aligned} &h_f = h_1 = h_2 , \quad E_x = E_{x1} = E_{x2} , \quad E_y = E_{y1} = E_{y2} , \\ &E_c = E_{cx} = E_{cy} \quad \text{and} \quad G_c = G_{cx} = G_{cy} . \end{aligned} \tag{A.27}$$

This simplifies the cross-section properties significantly as shown in Figure A.8.

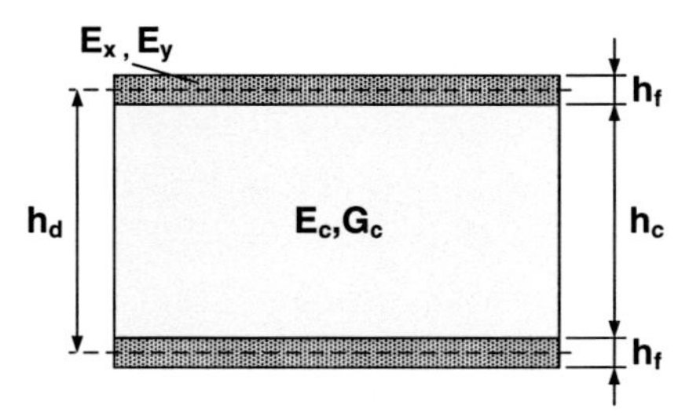

Figure A.8 – sandwich plate cross-section properties with identical orthotropic face sheets and isotropic core [Zen95]

The cross-sectional bending stiffness also simplifies and becomes similar to what has been described in section A.1 for the beam:

$$D_x = \frac{E_x h_f h_d^2}{2}, \qquad \text{and} \qquad D_y = \frac{E_y h_f h_d^2}{2}. \tag{A.28}$$

The torsional stiffness D_{xy} and the shear stiffnesses S_x and S_y of the sandwich plate can now be expressed using the thin face sheet and weak core assumptions as defined by Zenkert [Zen95] by:

$$\begin{aligned} S_x = S_y &= \frac{G_c\, d^2}{h_c}, \qquad \text{and} \\ D_{xy} &= \int 2z^2\, G_{xy} dz \approx \frac{2\, G_{xy1} h_1\, G_{xy2} h_2\, h_d^2}{G_{xy1} h_1 + G_{xy2} h_2} = G_{xy} h_f h_d^2 . \end{aligned} \tag{A.29}$$

The stresses and strains in a sandwich plate can be derived similar as shown for the beam. Thus the in-plane stresses due to a bending moment become [Zen95]

$$\sigma_x = \frac{M_x z_x E_x}{D_x}, \text{ and } \quad \sigma_y = \frac{M_y z_y E_y}{D_y}. \tag{A.30}$$

The stresses in the face sheets and the core due to a bending moment now become for a sandwich with thin face sheets and a weak core [Zen95]

$$\sigma_{fx} \approx \frac{M_x}{h_f h_d}, \quad \sigma_{fy} \approx \frac{M_y}{h_f h_d} \qquad \text{and} \qquad \sigma_c \approx 0 . \tag{A.31}$$

The stress strain relation for the face sheets now become according to Hooke's law [Zen95]

$$\varepsilon_x = \frac{\sigma_x}{E_x} - \nu_{yx} \frac{\sigma_y}{E_y} \qquad \text{and} \qquad \varepsilon_y = \frac{\sigma_y}{E_y} - \nu_{xy} \frac{\sigma_x}{E_x}. \tag{A.32}$$

For in-plane loads the strains can be calculated the same way as for the sandwich beam [Zen95]:

$$\varepsilon_{x0} = \frac{N_x}{2E_x h_f + E_c h_c} = \frac{N_x}{A_x} \qquad \text{and} \qquad \varepsilon_{y0} = \frac{N_y}{2E_y h_f + E_c h_c} = \frac{N_y}{A_y}. \tag{A.33}$$

The resulting stresses can be determined from equation (A.32). Strains and stresses due to bending and in-plane loads are superimposed. In-plane shear stresses due to torsional moments M_{xy} on the plate can be expressed by [Zen95]

$$\gamma_{xy} = \frac{2\, M_{xy}\, z}{D_{xy}} \quad \text{and} \quad \tau_{xy} = G_{xy} \gamma_{xy} . \tag{A.34}$$

For sandwich plates with thin face sheets and a weak core this separates into the following expressions for the face sheet and the core [Zen95]:

$$\gamma_{xy} = \frac{M_{xy}}{G_{f,xy}\, h_f\, h_d}, \qquad \tau_{fxy} \approx \frac{M_{xy}}{h_f\, h_d} \quad \text{and} \quad \tau_{cxy} \approx 0. \tag{A.35}$$

Zenkert [Zen95] calculates the core shear stresses due to transverse forces T_x and T_y in the same way as for the beam. Equation (A.21) is now applied separately for the sandwich x and y directions based on the thin face sheet and weak core assumption:

$$\tau_{c,xz} = \frac{T_x}{h_d}, \quad \tau_{c,yz} = \frac{T_y}{h_d}, \quad \gamma_{c,xz} = \frac{\tau_{c,xz}}{G_c} \quad \text{and} \quad \gamma_{c,yz} = \frac{\tau_{c,yz}}{G_c}. \tag{A.36}$$

A.3 Alternative notation for sandwich plate theory

Alternatively to the previously used sandwich notation a more general form based on the CLT may be used. The CLT applies Kirchhoff's plate theory and uses the plane strain stiffness matrix $[Q]$ for each layer of anisotropic material. Thus shear deformations are neglected and the problem is reduced to in-plane and bending loads that cause in- and out-of-plane deformations.

Using $[\overline{Q}]$ - the plane strain stiffness matrixes transformed into the global sandwich coordinate system – the sandwich laminate stiffness matrix $[A, B, D]$ can now be obtained by

$$\begin{aligned}[A, B, D] &= \int_0^h [\overline{Q}](1, z, z^2)dz \\ &= \sum_{k=1}^{N} [\overline{Q}]_k \left(z_k - z_{k-1}, \frac{1}{2}(z_k^2 - z_{k-1}^2), \frac{1}{3}(z_k^3 - z_{k-1}^3)\right).\end{aligned} \tag{A.37}$$

Here $[A]$ is the in-plane stiffness component, $[D]$ the bending stiffness component and $[B]$ describes the coupling of in- and out-of-plane deformations while z_k is the z-coordinate of the corresponding sandwich constituent. The elastic behavior of the laminate as a plate is now summarized in the expression

$$\begin{bmatrix} N_x \\ N_y \\ N_{xy} \\ M_x \\ M_y \\ M_{xy} \end{bmatrix} = \begin{bmatrix} A_{11} & A_{12} & A_{16} & B_{11} & B_{12} & B_{16} \\ & A_{22} & A_{26} & B_{12} & B_{22} & B_{26} \\ & & A_{66} & B_{16} & B_{26} & B_{66} \\ & & & D_{11} & D_{12} & D_{16} \\ & \text{sym.} & & & D_{22} & D_{26} \\ & & & & & D_{66} \end{bmatrix} \begin{bmatrix} \varepsilon_x \\ \varepsilon_y \\ \gamma_{xy} \\ \kappa_x \\ \kappa_y \\ \kappa_{xy} \end{bmatrix} \tag{A.38}$$

with: $\kappa_x = -\frac{\partial^2 w}{\partial x^2}$, $\kappa_y = -\frac{\partial^2 w}{\partial y^2}$ and $\kappa_{xy} = -\frac{\partial^2 w}{\partial x \, \partial y}$.

For a sandwich plate the CLT can be applied separately to the three components upper face sheet, core and lower face sheet. Assuming orthotropic and thin face sheets, the coupling and bending terms $[B]_{f1}$, $[B]_{f2}$, $[D]_{f1}$ and $[D]_{f2}$ of the face sheets become zero. As most core materials are at least orthotropic, the coupling term $[B]$ is equal to zero for all three components. Thus the components of the laminate stiffness matrix $[A, B, D]$ of a sandwich plate can be determined by the following expressions [Zen95]:

$$[A]_{sw} = [A]_{f1} + [A]_c + [A]_{f2} \,,$$
$$[B]_{sw} = \frac{h_c}{2}([A]_{f1} - [A]_{f2}) \quad \text{and} \tag{A.39}$$
$$[D]_{sw} = \frac{h_d^2}{4}([A]_{f1} + [A]_{f2}) + \frac{h_c^3}{12}[\bar{Q}]_c \,.$$

Here the coupling term $[B]$ of the sandwich plate arises only due to different face sheets properties. Thus using a fully symmetric sandwich with identical upper and lower face sheets leads to $[B]_{sw} = [0]$. It should finally be mentioned that in a fully symmetric sandwich made of orthotropic materials in the face sheets and the core the coupling between bending and twisting expressed by the terms $[D_{16}]_{sw}$ and $[D_{26}]_{sw}$ also becomes zero. In this case the bending stiffness terms can be connected to the previous notation by [Zen95]:

$$[D_{11}]_{sw} = \frac{D_x}{1-\nu_{xy}\,\nu_{yx}} \,, \qquad [D_{22}]_{sw} = \frac{D_y}{1-\nu_{xy}\,\nu_{yx}} \,,$$
$$[D_{12}]_{sw} = \frac{\nu_{yx} D_x}{1-\nu_{xy}\,\nu_{yx}} = \frac{\nu_{xy} D_y}{1-\nu_{xy}\,\nu_{yx}} \quad \text{and} \quad 2[D_{66}]_{sw} = D_{xy} \,. \tag{A.40}$$

The in-plane components $[A]_{sw}$ of the stiffness matrix can be calculated simply by summation. As mentioned in the beginning, the CLT is based on Kirchhoff's plate theory and thus neglects shear deformation and thus effectively assumes that the plate is reacting perfectly rigid on shear loads.

Due to the soft core of the sandwich this deformation mode can however not be neglected. Therefore Zenkert [Zen95] describes an approach from Whitney [Whi87] which defines the relation between transverse forces and out-of-plane shear strains similar to the laminate stiffness matrix applied in equation (A.35). This relation now becomes

$$\begin{bmatrix} T_y \\ T_x \end{bmatrix} = \begin{bmatrix} A_{44} & A_{45} \\ A_{45} & A_{55} \end{bmatrix} \begin{bmatrix} \gamma_{yz} \\ \gamma_{xz} \end{bmatrix} \quad \text{with} \quad A_{ij} = h_c C_{ij} \,. \tag{A.41}$$

Here the matrix C_{ij} is the anisotropic stiffness matrix of the core that directly couples the shear stresses τ_{xz} and τ_{yz} with the shear displacements γ_{xz} and γ_{yz}. As the face sheets are typically significantly more rigid to shear deformation than the core it is sufficient to use the core properties only and ignore the effect of the face sheets to the out-of-plane shear deformation. As most core materials are at least orthotropic this can be simplified and connected to the previous notation using the expressions [Zen95]

$$[A_{44}]_{sw} = h_c G_{c,yz} \approx S_y \,, \quad [A_{55}]_{sw} = h_c G_{c,xz} \approx S_x \quad \text{and} \quad [A_{45}]_{sw} = 0 \,. \tag{A.42}$$

A.4 Effective properties of orthotropic sandwich plates

As designers of composite sandwich structures want to capitalize on their ability to be tailored to the individual application, the sandwich face sheets are often orthotropic. In order to simplify the analytical description of the impact response, Olsson [Ols02] uses averaged properties. These are referred to as effective properties and described as follows

$$D^* = \sqrt{\frac{D_{11}D_{22}(\eta+1)}{2}}, \qquad \text{where} \qquad \eta = \frac{D_{12}+2D_{66}}{\sqrt{D_{11}D_{22}}}. \tag{A.43}$$

Here D^* is the effective plate stiffness. The components of the sandwich plate stiffness matrix D_{11}, D_{22}, D_{12} and D_{66} have been defined in equation (A.40). The effective shear stiffness S^* is defined as

$$S^* = \sqrt{A_{44}^* A_{55}^*} = \sqrt{K_{44}A_{44}K_{55}A_{55}}. \tag{A.44}$$

Here K_{44} and K_{55} are shear correction factors. Applying the typical sandwich theory assumptions of thin but orthotropic face sheets and a soft core, K_{44} and K_{55} are equal to 1 [Ols02]. A_{44} and A_{55} have been previously defined in equation (A.42).

Similarly the effective plate properties D_f^* and S_f^* of the face sheet only can be determined. Also the membrane properties of the face sheet are averaged by Olsson [Ols02]

$$E_f^* = \frac{1}{2\pi}\int_0^{2\pi}[E_f(\theta)]d\theta, \qquad \text{and} \qquad \nu_f^* = \frac{1}{2\pi}\int_0^{2\pi}[\nu_f(\theta)]d\theta. \tag{A.45}$$

Here E_f^* is the effective membrane stiffness of the face sheet and ν_f^* its Poisson ratio.

A.5 Foundation modulus K_z

For the foundation modulus K_z two solutions are described in equation (2.5) [Zen95]:

$$K_z = 0.28\, E_c \sqrt[3]{\frac{E_c}{D_f}} \quad \text{for} \quad h_f \ll h_c \quad \text{and}$$
$$K_z = \frac{E_c}{h_c} \quad \text{for} \quad h_f \approx h_c\,. \tag{A.46}$$

With the elastic material properties as described section B of the appendix the foundation modulus was calculated for CFRP face sheets and a PMI Rohacell RIST foam core with different densities. The results for 71RIST are summarized in Table A.1. Figure A.9 presents the results for different densities of the PMI foam.

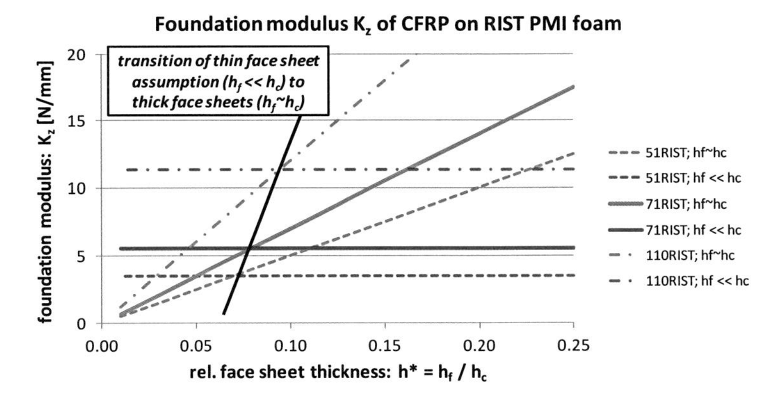

Figure A.9 – foundation modulus K_z as a function of the relative face sheet thickness h^* for Rohacell RIST foam: The black line displays the transition from thin face sheets ($h_f \ll h_c$) to thick face sheets ($h_f \approx h_c$)

Table A.1 – comparison of K_z for CFRP + Rohacell 71RIST and different geometries

h_f [mm]	h_c [mm]	h^* [-]	D_f [N*mm²]	E_c [MPa]	K_z ($t_f \ll t_c$) [N/mm]	K_z ($t_f \approx t_c$) [N/mm]
1.5	6	0.25	$1.6 \cdot 10^4$	105	17.5	5.5
1.5	10	0.15	$1.6 \cdot 10^4$	105	10.5	5.5
1.5	15	0.10	$1.6 \cdot 10^4$	105	7	5.5
1.5	25	0.06	$1.6 \cdot 10^4$	105	4.2	5.5
1.5	50	0.03	$1.6 \cdot 10^4$	105	2.1	5.5
1.5	150	0.01	$1.6 \cdot 10^4$	105	0.7	5.5

A.6 Non-dimensional sandwich properties

With the material properties as described section B of the appendix, non-dimensional properties of the sandwich can be determined for the creation of failure mode maps as summarized in equations (2.35). For a better illustration the definitions are repeated here:

$$h^* = h_f/h_c\,, \qquad L^* = L/h_c\,,$$
$$E^* = E_f/E_c\,, \qquad \sigma^* = \hat{\sigma}_f/\hat{\sigma}_c \qquad \text{and} \qquad (A.47)$$
$$S^* = \hat{\sigma}_c/\hat{\tau}_c\,.$$

Table A.2 – non-dimensional sandwich properties

Foam core	ρ [kg/m³]	h^* [-]	L^* [-]	E^* [-]	σ^* [-]	S^* [-]
51RIST	52	vertical axis	horizontal axis	876.9	619.2	1.05
71RIST	75	vertical axis	horizontal axis	534.8	297.7	1.26
110RIST	110	vertical axis	horizontal axis	318.9	138.4	1.53

B Appendix: Material properties and failure criteria

B.1 Fiber properties

Properties of the Toho Tenax HTS40 fiber applied [Ten12][Ikv04]:

Table B.1 – properties of Toho Tenax HTS carbon fibers

ρ [kg/m³]	E_{11} [MPa]	E_{22} [MPa]	G_{12} [MPa]	ν_{12} [-]	ν_{23} [-]	σ_{max} [MPa]	ε_{max} [%]
1.77	240000	15000	28500	0.2	0.5	4300	1.8

B.2 Matrix properties

Properties of the epoxy resin Hexcel RTM 6 used [Hex12]:

Table B.2 – properties of the epoxy resin Hexcel RTM 6

ρ [kg/m³]	E [MPa]	ν [-]	σ_{max} [MPa]	ε_{max} [%]
1.14	2890	0.3	75	3.4

B.3 Laminate properties

Intra- and interlaminar properties of the NCF used:

Table B.3 – elastic properties of the NCF UD ply

FVC [%]	ρ [g/cm³]	h_{ply} [mm]	E_{11} [MPa]	E_{22} [MPa]	G_{12} [MPa]	G_{23} [MPa]	ν_{12} [-]	ν_{23} [-]	$\alpha_{T,11}$ [K⁻¹]	$\alpha_{T,22}$ [K⁻¹]
58	1.51	0.132	126000	9100	4600	3800	0.28	0.20	0.4 x 10^{-6}	35 x 10^{-6}

Table B.4 – MAT54: strength and failure properties of the NCF UD ply

X_T [MPa]	X_C [MPa]	Y_T [MPa]	Y_C [MPa]	S_L [MPa]	ε_{XT} [%]	ε_{XC} [%]	ε_Y [%]	ε_{SL} [%]
2300	1400	85	200	125	2.8	3.4	20	20

Table B.5 – MAT262: Strength properties of the NCF UD ply

X_T [MPa]	X_C [MPa]	Y_T [MPa]	Y_C [MPa]	S_L [MPa]	X_{T0} [MPa]	X_{C0} [MPa]	$\hat{\tau}_y$ [MPa]	G_{tan} [MPa]
2300	1400	80	200	125	200	150	100	180

Table B.6 – MAT262: Fracture mechanics properties of the NCF UD ply

G_{XT} [kJ/m²]	G_{XT0} [kJ/m²]	G_{XC} [kJ/m²]	G_{XC0} [kJ/m²]	G_{YT} [kJ/m²]	G_{YC} [kJ/m²]	G_{SL} [kJ/m²]
100	25	90	20	2.1	7.5	7.5

Table B.7 – CFRP interface properties used for the cohesive zone model

l_{cz} [mm]	h_{cz} [mm]	K_N [MPa/mm]	K_T [MPa/mm]	G_{Ic} [J/m²]	G_{IIc} [J/m²]	τ_I^0 [MPa]	τ_{II}^0 [MPa]
3.0	0.001	7.0*10⁴	3.2*10⁴	426	1500	33	42

B.4 MAT262: Constitutive law with damage

The continuum damage model of Maimí, Camanho et al. [Mai07,Mai07a] implemented into LS-DYNA as material model #262 (MAT262) [Lst13a] is based on the thermodynamics of irreversible processes. The complementary free energy density G^{free} of a ply within the laminate is defined as

$$G^{free} = \frac{\sigma_{11}^2}{2(1-\psi_1)E_{11}} + \frac{\sigma_{22}^2}{2(1-\psi_2)E_{22}} + \frac{\sigma_{12}^2}{2(1-\psi_6)G_{12}} - \frac{\nu_{12}}{E_{11}}\sigma_{11}\sigma_{22} \,. \tag{B.1}$$

For simplicity thermal and moisture effects are omitted here. The variables ψ_1, ψ_2 and ψ_6 are state variables of the material and describe the effect of damage on the individual material stiffnesses. The second derivative of G^{free} with respect to the stress tensor leads to the compliance tensor [S] which is thus defined as

$$[S] = \frac{\partial^2 G^{free}}{\partial \sigma^2} = \begin{bmatrix} \frac{1}{(1-\psi_1)E_{11}} & -\frac{\nu_{21}}{E_{22}} & 0 \\ -\frac{\nu_{12}}{E_{11}} & \frac{1}{(1-\psi_2)E_{22}} & 0 \\ 0 & 0 & \frac{1}{(1-\psi_6)G_{12}} \end{bmatrix} . \tag{B.2}$$

The damage variables ψ_1 and ψ_2 describe material failure in the longitudinal (fiber) direction and the transverse (matrix) direction. The damage variable ψ_6 is associated with the degradation of the in-plane shear stiffness which is predominantly caused by matrix cracking. The model also tracks damages causes by tensile loads (ψ_+) and compressive loads

(ψ_-) separately to describe the effect of crack closure in the fiber and matrix direction during compressive loads. The variables ψ_1 and ψ_2 are now defined by

$$\psi_1 = \psi_{1+}\frac{\langle\sigma_{11}\rangle}{|\sigma_{11}|} + \psi_{1-}\frac{\langle-\sigma_{11}\rangle}{|\sigma_{11}|} \qquad \text{and} \tag{B.3}$$

$$\psi_2 = \psi_{2+}\frac{\langle\sigma_{22}\rangle}{|\sigma_{22}|} + \psi_{2-}\frac{\langle-\sigma_{22}\rangle}{|\sigma_{22}|} \qquad \text{where} \qquad \langle x\rangle := (x + |x|)/2\ . \tag{B.4}$$

From this four damage activation functions F_N are defined by

$$F_{1+} = \phi_{1+} - r_{1+} \leq 0\ , \qquad F_{1-} = \phi_{1-} - r_{1-} \leq 0\ , \tag{B.5}$$

$$F_{2+} = \phi_{2+} - r_{2+} \leq 0 \qquad \text{and} \qquad F_{2-} = \phi_{2-} - r_{2-} \leq 0\ . \tag{B.6}$$

Here ϕ_N are the failure criteria from equations (5.15) to (5.18) while r_N describes the state of the elastic threshold. The value of r_N is in the undamaged state 1 and increases as damage occurs. In this model the failure criteria ϕ_N are computed based on the effective stresses $\tilde{\sigma}$ as

$$\{\tilde{\sigma}\} = [S_0]^{-1}\{\varepsilon\}\ . \tag{B.7}$$

Here $[S_0]$ is the undamaged compliance tensor which can be obtained from equation (B.2) by inserting $\psi_1 = \psi_2 = \psi_6 = 0$. The elastic thresholds r_N are now defined as the maximum value of ϕ_N that has occurred during the complete load history of the material. For longitudinal (fiber) failure this reads

$$r_{1+} = \max\{1, \max_{s=0,t}\{\phi_{1+}^s\}, \max_{s=0,t}\{\phi_{1-}^s\}\} \qquad \text{and} \tag{B.8}$$

$$r_{1-} = \max\{1, \max_{s=0,t}\{\phi_{1-}^s\}\}\ . \tag{B.9}$$

Here interaction of the failure modes is included as tensile fiber cracks close during compressive loading while compressive fiber failure initiated by kink bands does not. Crack closure during compressive loading is typically not perfect and the reduction factor $C_{A1}^{\pm}$ for the compressive stiffness is used which is defined by

$$C_{A1}^{\pm} \approx C_B\frac{V_f E_f}{V_m E_m + V_f E_f} \approx C_B\frac{E_{11} - E_{22}}{E_{22}}\ . \tag{B.10}$$

Here C_B is an adjustment parameter which anticipates that for $C_B = 1$ stiffness recovery is during compressive loading is perfect while $C_B = 0$ assumes that stiffness recovery is limited to matrix closure.

For transverse (matrix) failure the definition of the elastic threshold reads similar as matrix cracks generated during tensile loading occur typically perpendicular to the laminate plane and thus close perfectly during compressive loading. Cracks that occur during compressive loading are typically inclined at an angle of $\alpha_0 = 53 \pm 2°$. During tensile loading this crack can not anymore transfer loads and is thus as effective as a crack perpendicular to the lamina. The elastic thresholds are thus defined by

$$r_{2+} = \max\{1, \max_{s=0,t}\{\phi_{2+}^{s}\}, \max_{s=0,t}\{\phi_{2-}^{s}\}\} \quad \text{and} \tag{B.11}$$

$$r_{2-} = \max\{1, \max_{s=0,t}\{\phi_{2-}^{s}\}\} \; . \tag{B.12}$$

Damage evolution is now controlled by the fracture toughness comparable to the failure model for interlaminar failure described in section 5.2. Maimí, Camanho et al. [Mai07a] propose linear-exponential damage evolution for longitudinal (fiber) failure and exponential damage evolution for transverse (matrix) failure. The implementation into LS-DYNA applies however a simpler bilinear damage evolution for fiber failure and linear damage evolution for matrix failure. Additionally a one dimensional elasto-plastic formulation is applied to account for the characteristic in-plane shear behavior of fiber reinforced plastics. The resulting material laws are shown in principal in Figure B.1.

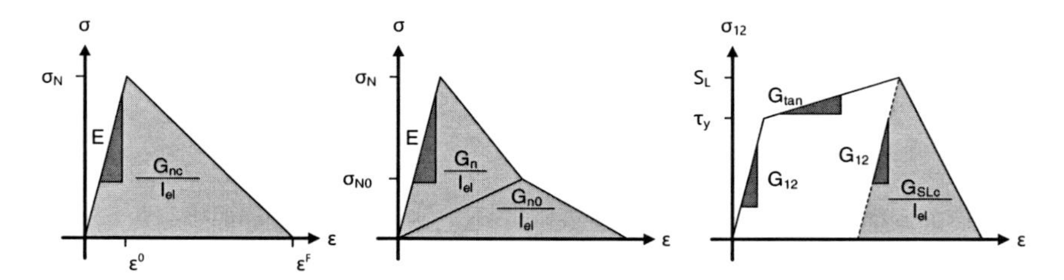

Figure B.1 – material laws applied in MAT262: Linear damage evolution (left), bi-linear damage evolution (center) and elasto-plastic behavior coupled with linear damage evolution (right)

As no exact definition of the linear and bilinear damage functions is provided in the LS-DYNA handbook [Lst13a], only the principal dependence of the damage variables ψ_N used for degradation of the stiffnesses on the material state variables r_N can be given here with

$$\psi_{1+} = 1 - f(E_{11}, X_T, G_{XT}, X_{T0}, G_{XT0}, r_{1+}) \, , \tag{B.13}$$

$$\psi_{1-} = 1 - f(E_{11}, X_C, G_{XC}, X_{C0}, G_{XC0}, r_{1-}) \, , \tag{B.14}$$

$$\psi_{2+} = 1 - f(E_{22}, Y_T, G_{YT}, r_{2+}) \, , \tag{B.15}$$

$$\psi_{2-} = 1 - f(E_{22}, Y_C, G_{YC}, r_{2-}) \quad \text{and} \tag{B.16}$$

$$\psi_6 = 1 - f(G_{12}, S_L, G_{SL}, r_{2+}) \; . \tag{B.17}$$

Here $f(\,)$ is the damage function and l_{el} the characteristic element length. It is noted that ψ_6 is defined as dependent on the material state variable r_{2+} which is associated with transverse (matrix) cracks as shear loads are already taken into account with the corresponding failure criterion ϕ_{2+}. From this it becomes clear that a maximum element size can be defined for each damage mode N by

$$l_{el,max} = \frac{2\,G_{nc}}{\sigma_N\,\varepsilon^0} \quad \text{with} \tag{B.18}$$

$$\varepsilon^0 = E_N\,\sigma_N\ .$$

In case of $l_{el,max}$ failure will take place very abruptly as the complete fracture energy is already consumed as the element reaches its material strength σ_N. Thus in practice significantly smaller element sizes have to be used for a proper description of a continuous failure behavior as the cohesive zone describing crack propagation should stretch across 2 – 3 elements. The resulting minimum element size for a cohesive zone stretching across n_{el}^0 elements can be determined using the recommendations of Turon et al. [Tur07] from equation (5.21). Solving this for the element length l_{el} provides the maximum element size

$$l_{el,max} = \frac{9\,\pi\,E\,G_{nc}}{32\,n_{el}^0\,\sigma_N}\ . \tag{B.19}$$

B.5 Properties of the Rohacell RIST PMI foam

Table B.8 – properties of the PMI foam Rohacell RIST [Evo13]

	ρ [kg/m³]	E_c [MPa]	G_c [MPa]	$\hat{\sigma}_{c,T}$ [MPa]	$\hat{\sigma}_{c,C}$ [MPa]	$\hat{\tau}_c$ [MPa]	ε_{max} [%]	α_T [K⁻¹]
51RIST	52	75	24	1.6	0.8	0.8	3.0	35×10^{-6}
71RIST	75	105	42	2.2	1.7	1.3	3.0	35×10^{-6}
110RIST	110	180	70	3.7	3.6	2.4	3.0	35×10^{-6}

Curve fits of the above values express the material properties of the PMI foam relative to their density by

$$E_c = 4500\ \text{MPa}\left(\frac{\rho_c}{\rho_u}\right)^{1.35},$$
$$G_c = 1750\ \text{MPa}\left(\frac{\rho_c}{\rho_u}\right)^{1.35} \quad \text{and} \tag{B.20}$$
$$\hat{\tau}_c = 85\ \text{MPa}\left(\frac{\rho_c}{\rho_u}\right)^{1.5}.$$

Table B.9 – calibrated material properties of the PMI foam Rohacell 71RIST

ρ [kg/m³]	E_c [MPa]	G_c [MPa]	$\hat{\sigma}_{c,T}$ [MPa]	$\hat{\sigma}_{c,C}$ [MPa]	$\hat{\tau}_c$ [MPa]	ε_{lu} [%]	α_T [K⁻¹]
75	105	42	2.2	1.85	1.4	65	35×10^{-6}

B.6 Invariant failure criterion for PMI foams

The invariant failure criterion for PMI foams of Kraatz is based on an ellipsoidal potential [Kra07][Rot08]. This potential Γ is defined as

$$\Gamma = \frac{3\,J_2' + a_1 \sigma_V\, J_1 + a_2 J_1^2}{1 + a_1 + a_2} = \sigma_V^2 \,. \tag{B.21}$$

Here σ_V is the equivalent stress used for comparison while J_1 and J_2' are the invariants. These are defined as

$$\begin{aligned} J_1 &= \sigma_{11} + \sigma_{22} + \sigma_{33} \qquad \text{and} \\ J_2' &= \frac{1}{3}\begin{bmatrix} \sigma_{11}^2 + \sigma_{22}^2 + \sigma_{33}^2 - \sigma_{11}\sigma_{22} - \sigma_{11}\sigma_{33} - \sigma_{22}\sigma_{33} \\ +3(\sigma_{12}^2 + \sigma_{23}^2 + \sigma_{13}^2) \end{bmatrix} . \end{aligned} \tag{B.22}$$

The equivalent stress σ_V is determined to

$$\sigma_V = \frac{\sqrt{(12a_2 + 12a_1 + 12)\, J_2' + (4a_2^2 + (4a_1 + 4)a_2 + a_1^2) J_1^2} + a_1 J_1}{2a_2 + 2a_1 + 2} \,. \tag{B.23}$$

The material effort f_E can now be calculated by comparing the equivalent stress with the tensile yield strength $\hat{\sigma}_{c,T}$ of the PMI foam by

$$f_E = \frac{\sigma_V}{\hat{\sigma}_{c,T}} \,. \tag{B.24}$$

Failure due to yield or fracture of the foam occurs for values of $f_E > 1$. The missing material parameters a_1 and a_2 are defined by Kraatz to

$$a_1 = \frac{{b_{V,s}}^2 (b_{V,C} - 1)}{b_{V,C}} \qquad \text{and} \qquad a_2 = \frac{{b_{V,s}}^2}{b_{V,C}} - 1 \,. \tag{B.25}$$

Here $b_{V,s}$ is the ratio of the foam core shear strength $\hat{\tau}_c$ to the tensile strength $\hat{\sigma}_{c,T}$ and $b_{V,C}$ the ratio of the compressive strength $\hat{\sigma}_{c,C}$ to the tensile strength $\hat{\sigma}_{c,T}$. This leads to the following expressions for $b_{V,s}$ and $b_{V,C}$:

$$b_{V,s} = \sqrt{3}\,\frac{\hat{\tau}_c}{\hat{\sigma}_{c,T}} \qquad \text{and} \qquad b_{V,C} = \frac{\hat{\sigma}_{c,C}}{\hat{\sigma}_{c,T}} \,. \tag{B.26}$$

The strength ratios and the material parameters of the Rohacell 71RIST foam are summarized in Table B.10 based on the material properties from Table B.8.

Table B.10 – properties of the Rohacell RIST PMI foam

$b_{V,s}$ [-]	$b_{V,C}$ [-]	a_1 [-]	a_2 [-]
1.0235	0.7727	-0.3081	0.3556

B.7 Rate sensitive material properties

Table B.11 – MAT54: Strain rate dependent scaling of the strength properties of the NCF UD ply

$\dot{\varepsilon}$ [s^{-1}]	X_T [MPa]	X_C [MPa]	Y_T [MPa]	Y_C [MPa]	S_L [MPa]
10^{-2}	2300	1400	80	200	125
10^{-1}	2300	1400	80	200	125
1	2300	1400	80	200	137.5
10^{1}	2300	1540	96	240	150
10^{2}	2300	1680	112	280	167.5
10^{3}	2300	1820	128	320	175
10^{4}	2300	1960	128	320	175
10^{5}	2300	1960	128	320	175

Table B.12 – strain rate dependent scaling of strength properties of Rohacell 71RIST

$\dot{\varepsilon}$ [s^{-1}]	E_c [MPa]	G_C [MPa]	$\hat{\sigma}_{c,T}$ [MPa]	$\hat{\sigma}_{c,C}$ [MPa]	$\hat{\tau}_C$ [MPa]
10^{-4}	105	42	2.20	1.80	1.40
10^{-3}	105	42	2.20	1.80	1.40
10^{-2}	105	42	2.35	1.93	1.50
10^{-1}	105	42	2.51	2.05	1.60
1	105	42	2.66	2.18	1.69
10^{2}	105	42	2.66	2.18	1.69
10^{4}	105	42	2.66	2.18	1.69

B.8 Properties of the sandwich interface

Table B.13 – material properties of the sandwich interface

h_{intf} [mm]	K_N [MPa/mm]	K_T [MPa/mm]	G_{Ic} [J/m^2]	G_{IIc} [J/m^2]	τ_I^0 [MPa]	τ_{II}^0 [MPa]
0.2	1.2*10^4	0.45*10^4	150	150	2.2	1.4

C Appendix: Test results

C.1 RT impact tests

RT tests		h_{core} = 6.5 mm	h_{core} = 10.0 mm	h_{core} = 16.3 mm	h_{core} = 25.7 mm	h_{core} = 35.5 mm	
2 NCF plies	h_{face} = 0.75 mm	H1a	H1b	H1c	-	-	(3 heights)
4 NCF plies	h_{face} = 1.5 mm	6.5	10.0	16.3	25.7	35.5	(5 heights)
6 NCF plies	h_{face} = 2.25 mm	-	-	H2a	H2b	H2c	(3 heights)
8 NCF plies	h_{face} = 3.0 mm	-	-	H3a	H3a	H3b	(3 heights)
Specimen count		6	6	12	9	9	42

Figure C.1 – RT impact tests: Test matrix and naming convention

Table C.1 – RT impact tests: Details of tested sandwich configurations

confi-guration	NCF plies	h_{face} [mm]	stacking sequence	h_{core} [mm]	foam type	E_{imp} [J]	specimens
RT_H1a	2	0.75	$[(45/0/135)_s/$ core$/(135/0/45)_s]$	6.5	71RIST	12 / 20 / 35	3
RT_H1b	2	0.75	see RT_H1a	10.0	71RIST	12 / 20 / 35	3
RT_H1c	2	0.75	see RT_H1a	16.3	71RIST	20 / 35 / 35	3
RT_6.5	4	1.50	$[((45/0/135)_s)_2/$ core$/((135/0/45)_s)_2]$	6.5	71RIST	12 / 20 / 35	3
RT_10.0	4	1.50	see RT_6.5	10.0	71RIST	12 / 20 / 35	3
RT_16.3	4	1.50	see RT_6.5	16.3	71RIST	20 / 35 / 50	3
RT_25.7	4	1.50	see RT_6.5	25.7	71RIST	20 / 35 / 50	3
RT_35.5	4	1.50	see RT_6.5	35.5	71RIST	35 / 50 / 90	3
RT_H2a	6	2.25	$[((45/0/135)_s)_3/$ core$/((135/0/45)_s)_3]$	16.3	71RIST	35 / 35 / 50	3
RT_H2b	6	2.25	see RT_H2a	25.7	71RIST	35 / 50 / 90	3
RT_H2c	6	2.25	see RT_H2a	35.5	71RIST	35 / 50 / -	2 (3)
RT_H3a	8	3.00	$[((45/0/135)_s)_4/$ core$/((135/0/45)_s)_4]$	16.3	71RIST	35 / 35 / 50	3
RT_H3b	8	3.00	see RT_H3a	25.7	71RIST	35 / 50 / 90	3
RT_H3c	8	3.00	see RT_H3a	35.5	71RIST	35 / 50 / 90	3

Table C.2 – RT impact tests: Test results of specimens with 2-ply face sheets (0.75 mm thick)

Configuration / Panel:	core thickn. [mm]	face sheet thickn. [mm]	planed impact energy [J]	applied impact energy [J]	peak force [N]	max. displacement of impactor [mm]	dent depth [mm]	dissipated energy [J]	dissipated energy [%]	damage size 0°-direction [mm]	damage size 90°-direction [mm]	planar damage diameter [mm]	damage area [mm²]	core shear crack
RT_H1a_P01	6.5	0.75	20	20.04	2549	13.14	2.65	15.02	75	156	138	147	21528	x
RT_H1a_P02	6.5	0.75	35	34.98	3894	17.56	3.53	21.64	62	268	205	237	54940	x
RT_H1a_P03	6.5	0.75	12	12.05	2732	9.11	0.80	3.88	32	135	103	119	13905	x
RT_H1b_P01	10	0.75	20	20.04	2605	12.09	3.63	13.71	68	160	135	148	21600	x
RT_H1b_P02	10	0.75	35	34.98	3027	17.73	6.60	27.37	78	210	175	193	36750	x
RT_H1b_P03	10	0.75	12	12.05	1978	8.19	1.75	7.67	64	126	106	116	13356	x
RT_H1c_P01	16.3	0.75	20	20.04	3083	10.10	4.70	15.09	75	34	32	33	1088	-
RT_H1c_P02	16.3	0.75	35	34.98	2948	16.16	13.28	33.36	95	108	92	100	9936	-
RT_H1c_P03	16.3	0.75	35	34.98	3041	15.83	12.03	31.90	91	100	83	92	8300	-

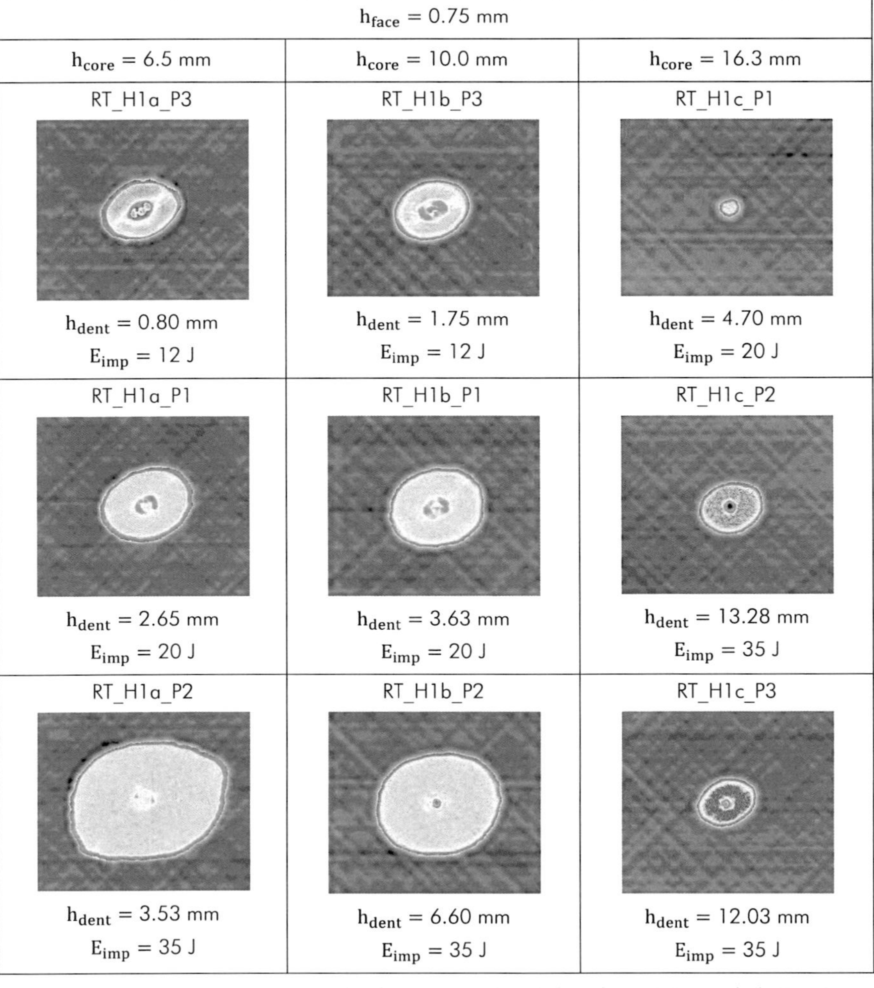

Figure C.2 – RT impact tests: NDI results of specimens with 2-ply face sheets (0.75 mm thick) [Fri11]

Table C.3 – RT impact tests: Test results of specimens with 4-ply face sheets (1.5 mm thick)

Configuration / Panel:	core thickn. [mm]	face sheet thickn. [mm]	planed impact energy [J]	applied impact energy [J]	peak force [N]	max. displacement of impactor [mm]	dent depth [mm]	dissipated energy [J]	dissipated energy [%]	damage size 0°-direction [mm]	damage size 90°-direction [mm]	planar damage diameter [mm]	damage area [mm²]	core shear crack
RT_6,5_P01	6.5	1.5	20	20.04	4030	9.08	0.10	7.66	38	180	142	161	25560	x
RT_6,5_P02	6.5	1.5	35	34.98	4536	12.47	0.91	21.12	60	272	224	248	60928	x
RT_6,5_P03	6.5	1.5	12	12.05	3102	6.71	0.25	4.71	39	114	92	103	10488	x
RT_10,0_P01	10	1.5	20	20.04	3758	8.52	0.25	9.30	46	225	185	205	41625	x
RT_10,0_P02	10	1.5	35	34.98	4058	12.15	0.95	22.92	66	300	250	275	75000	x
RT_10,0_P03	10	1.5	12	12.05	3707	5.88	0.15	4.28	35	25	18	22	450	-
RT_16,3_P01	16.3	1.5	20	20.04	5731	6.67	0.13	9.87	49	30	26	28	780	-
RT_16,3_P02	16.3	1.5	35	34.98	5239	10.27	1.85	27.08	77	36	32	34	1152	-
RT_16,3_P03	16.3	1.5	50	53.52	5694	12.91	3.23	39.73	74	40	36	38	1440	-
RT_25,7_P01	25.7	1.5	20	20.04	6218	6.07	1.00	16.95	85	38	34	36	1292	-
RT_25,7_P02	25.7	1.5	35	34.98	5239	10.03	3.11	30.31	87	42	36	39	1512	-
RT_25,7_P03	25.7	1.5	50	53.52	5038	12.96	6.54	41.96	78	42	40	41	1680	-
RT_35,5_P01	35.5	1.5	35	34.98	6026	9.28	2.86	30.35	87	34	32	33	1088	-
RT_35,5_P02	35.5	1.5	50	53.52	5642	12.03	6.68	42.40	79	40	38	39	1520	-
RT_35,5_P03	35.5	1.5	90	94.42	5155	25.73	24.20	86.36	91	42	48	45	2016	-

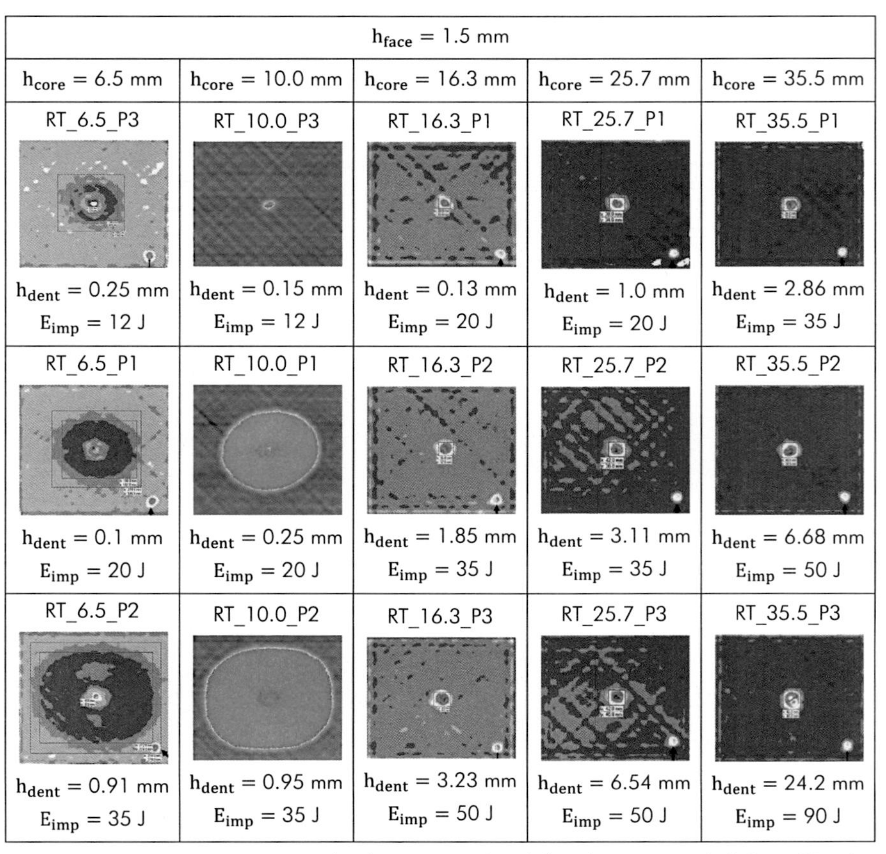

Figure C.3 – RT impact tests: NDI results of specimens with 4-ply face sheet (1.5 mm thick) [Bie11][Fri11]

Table C.4 – RT impact tests: Test results of specimens with 6-ply face sheets (2.25 mm thick)

Configuration / Panel:	core thickn. [mm]	face sheet thickn. [mm]	planed impact energy [J]	applied impact energy [J]	peak force [N]	max. dis-placement of impactor [mm]	dent depth [mm]	dissipated energy [J]	dissipated energy [%]	damage size 0°-direction [mm]	damage size 90°-direction [mm]	planar damage diameter [mm]	damage area [mm²]	core shear crack
RT_H2a_P01	16.3	2.25	35	34.98	7634	7.72	1.00	21.98	63	36	30	33	1080	-
RT_H2a_P02	16.3	2.25	50	53.52	7859	10.79	0.78	35.05	65	335	295	315	98825	x
RT_H2a_P03	16.3	2.25	35	34.98	7976	7.78	1.10	23.75	68	45	33	39	1485	-
RT_H2b_P01	25.7	2.25	35	34.98	8458	7.35	1.30	26.45	76	40	33	37	1320	-
RT_H2b_P02	25.7	2.25	50	53.52	8098	9.40	2.48	37.30	70	45	38	42	1710	-
RT_H2b_P03	25.7	2.25	90	94.42	7812	15.68	6.25	68.73	73	48	38	43	1824	-
RT_H2c_P01	35.5	2.25	35	34.98	8196	7.29	1.63	27.20	78	34	32	33	1088	-
RT_H2c_P02	35.5	2.25	50	53.52	8060	9.26	2.58	37.05	69	38	34	36	1292	-
RT_H2c_P03	35.5	2.25	-	*Low temperature test with fixed support: cf. Frost tests										

h_{face} = 2.25 mm		
h_{core} = 16.3 mm	h_{core} = 25.7 mm	h_{core} = 35.5 mm
RT_H2a_P1	RT_H2b_P1	RT_H2c_P1
h_{dent} = 1.00 mm E_{imp} = 35 J	h_{dent} = 1.30 mm E_{imp} = 35 J	h_{dent} = 1.63 mm E_{imp} = 35 J
RT_H2a_P3	RT_H2b_P2	RT_H2c_P2
h_{dent} = 1.10 mm E_{imp} = 35 J	h_{dent} = 2.48 mm E_{imp} = 50 J	h_{dent} = 2.58 mm E_{imp} = 50 J
RT_H2a_P2	RT_H2b_P3	RT_H2c_P3 (-55 °C)
h_{dent} = 0.78 mm E_{imp} = 50 J	h_{dent} = 6.25 mm E_{imp} = 90 J	see Frost test series

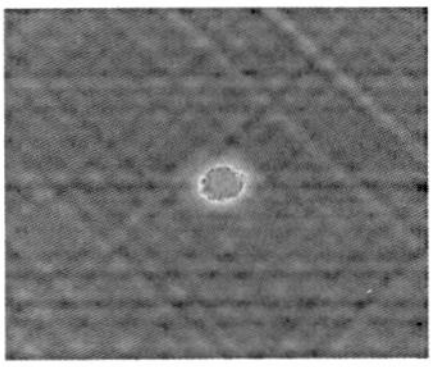

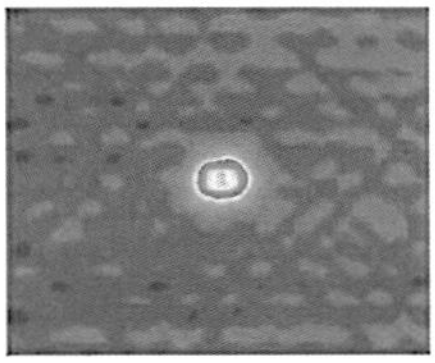

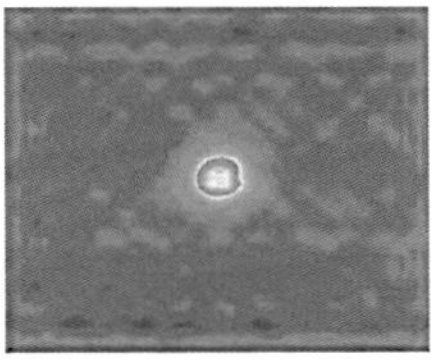

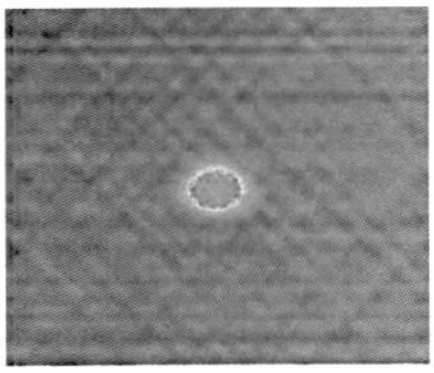

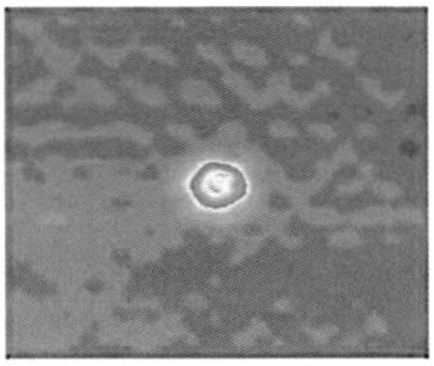

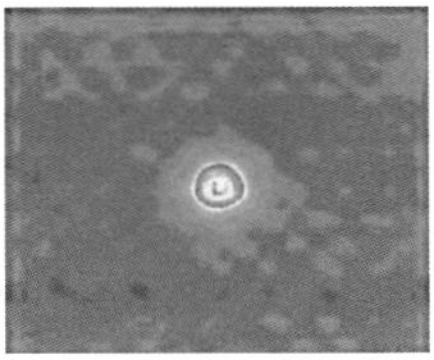

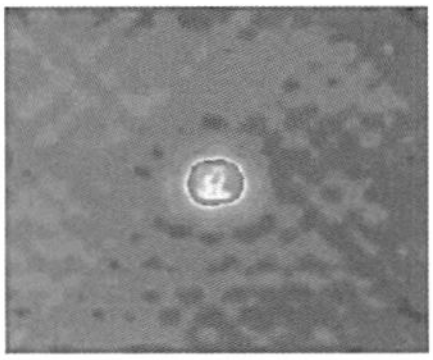

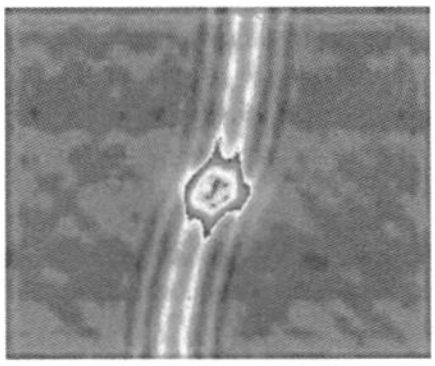

Figure C.4 – RT impact tests: NDI results of specimens with 6-ply face sheets (2.25 mm thick) [Fri11]

Table C.5 – RT impact tests: Test results of specimens with 8-ply face sheets (3.0 mm thick)

Configuration / Panel:	core thickn. [mm]	face sheet thickn. [mm]	planed impact energy [J]	applied impact energy [J]	peak force [N]	max. displacement of impactor [mm]	dent depth [mm]	dissipated energy [J]	dissipated energy [%]	damage size 0°-direction [mm]	damage size 90°-direction [mm]	planar damage diameter [mm]	damage area [mm²]	core shear crack
RT_H3a_P01	16.3	3	35	34.98	9897	6.34	0.25	18.72	54	34	26	30	884	-
RT_H3a_P02	16.3	3	50	53.52	11715	7.55	0.28	39.04	73	302	248	275	74896	x
RT_H3a_P03	16.3	3	35	34.98	10075	6.53	0.15	17.20	49	10	10	10	100	-
RT_H3b_P01	25.7	3	35	34.98	10277	6.31	0.20	18.90	54	41	30	36	1230	-
RT_H3b_P02	25.7	3	50	53.52	10609	7.50	1.13	40.42	76	43	37	40	1591	-
RT_H3b_P03	25.7	3	90	94.42	11284	11.42	2.05	64.80	69	49	42	46	2058	-
RT_H3c_P01	35.5	3	35	34.98	11181	6.08	0.18	21.17	61	36	30	33	1080	-
RT_H3c_P02	35.5	3	50	53.52	11298	7.40	1.18	38.48	72	42	36	39	1512	-
RT_H3c_P03	35.5	3	90	94.42	11139	11.73	2.70	64.91	69	49	43	46	2107	-

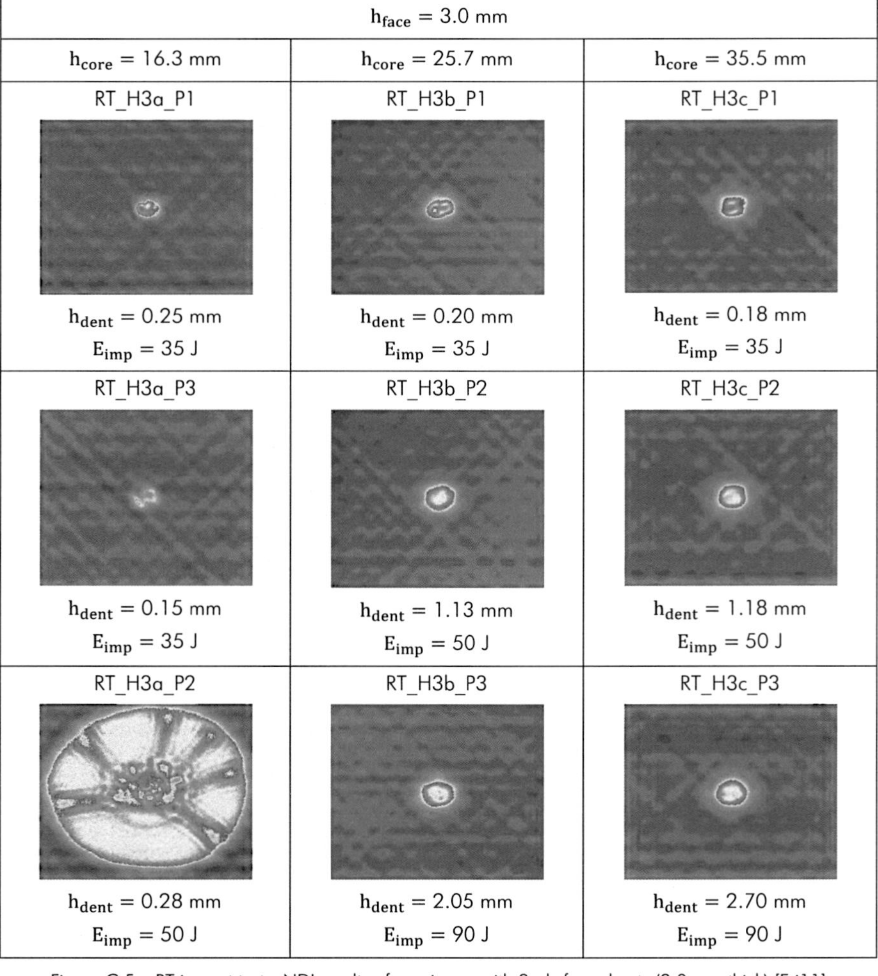

Figure C.5 – RT impact tests: NDI results of specimens with 8-ply face sheets (3.0 mm thick) [Fri11]

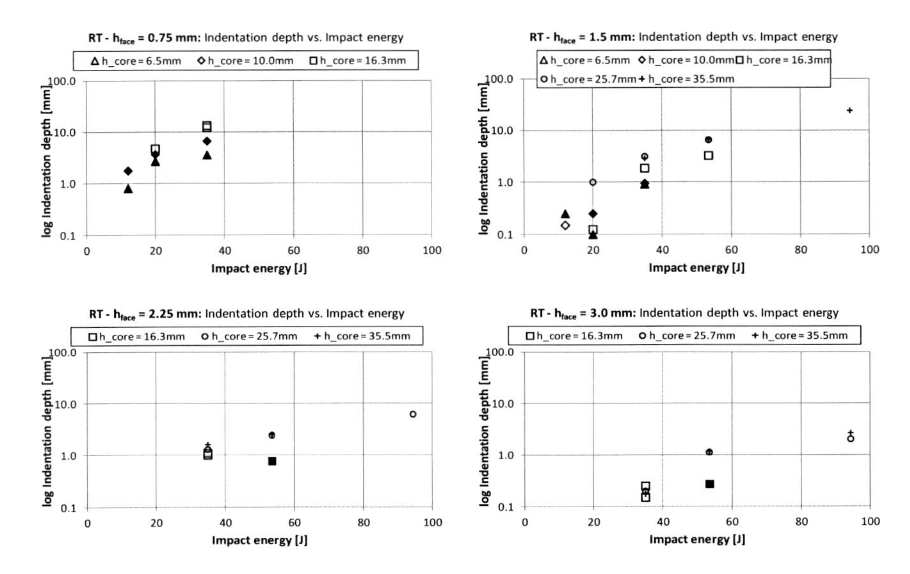

Figure C.6 – RT impact tests: Indentation depth vs. impact energy sorted by face sheet and core thicknesses; empty symbols: Face sheet rupture, filled symbols: Core shear failure.

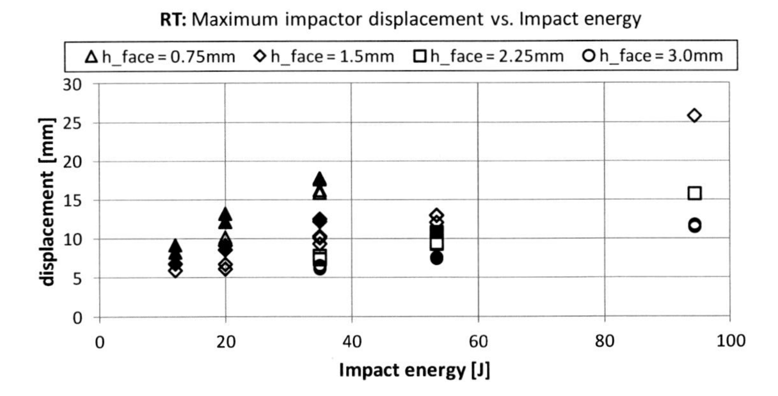

Figure C.7 – RT impact tests: Maximum impactor displacement vs. impact energy; empty symbols: Face sheet rupture, filled symbols: Core shear failure

C.2 Frost (-55°C) impact tests

Frost tests		h_{core} = 6.5 mm	h_{core} = 10.0 mm	h_{core} = 16.3 mm	h_{core} = 25.7 mm	h_{core} = 35.5 mm	
2 NCF plies	h_{face} = 0.75 mm	-	-	-	-	-	(0 heights)
4 NCF plies	h_{face} = 1.5 mm	-	-	-	25.7-1.5	35.5-1.5	(2 heights)
6 NCF plies	h_{face} = 2.25 mm	-	-	-	25.7-2.25	35.5-2.25	(2 heights)
8 NCF plies	h_{face} = 3.0 mm	-	-	-	-	-	(0 heights)
Specimen count		0	0	0	6	6	12

Figure C.8 – Frost impact tests: Test matrix and naming convention

Table C.6 – Frost impact tests: Details of tested sandwich configurations

confi-guration	NCF plies	h_{face} [mm]	stacking sequence	h_{core} [mm]	foam type	E_{imp} [J]	specimens
Frost_ 25.7-1.5	4	1.50	$[((45/0/135)_s)_2/$ core$/((135/0/45)_s)_2]$	25.7	71RIST	20 / 35 / 50	3
Frost_ 35.5-1.5	4	1.50	$[((45/0/135)_s)_2/$ core$/((135/0/45)_s)_2]$	35.5	71RIST	20 / 35 / 50	3
Frost_ 25.7-2.25	6	2.25	$[((45/0/135)_s)_3/$ core$/((135/0/45)_s)_3]$	25.7	71RIST	20 / 35 / 50	3
Frost_ 35.5-2.25	6	2.25	$[((45/0/135)_s)_3/$ core$/((135/0/45)_s)_3]$	35.5	71RIST	20 / 35 / 50 +50 (RT_H2c_P3)	3 (4)

Table C.7 – Frost impact tests: Test results

Configuration / Panel:	core thickn. [mm]	face sheet thickn. [mm]	planed impact energy [J]	applied impact energy [J]	peak force [N]	max. dis-placement of impactor [mm]	dent depth [mm]	dissipated energy [J]	dissipated energy [%]	damage size 0°-direction [mm]	damage size 90°-direction [mm]	planar damage diameter [mm]	damage area [mm²]	core shear crack	thermal foam crack
FROST_25,7-1,5_P01	25.7	1.5	20	20.04	4569	7.08	1.28	14.94	75	48	40	44	1920	-	-
FROST_25,7-1,5_P02	25.7	1.5	35	34.98	5127	10.97	3.00	29.92	86	54	62	58	3348	-	x
FROST_25,7-1,5_P03	25.7	1.5	50	53.52	6045	13.74	8.63	42.37	79	60	66	63	3960	-	x
FROST_25,7-2,25_P01	25.7	2.25	20	20.04	6954	5.34	0.18	11.41	57	54	46	50	2484	-	-
FROST_25,7-2,25_P02	25.7	2.25	35	34.98	7399	7.63	1.73	26.13	75	60	52	56	3120	-	-
FROST_25,7-2,25_P03	25.7	2.25	50	53.52	7868	10.68	0.78	35.33	66	340	320	330	108800	x	-
FROST_35,5-1,5_P01	35.5	1.5	20	20.04	6218	6.13	1.05	15.08	75	50	42	46	2100	-	-
FROST_35,5-1,5_P03	35.5	1.5	35	34.98	5272	9.97	5.48	29.02	83	44	50	47	2200	-	x
FROST_35,5-1,5_P02	35.5	1.5	50	53.52	5394	13.53	6.68	43.29	81	42	58	50	2436	-	x
FROST_35,5-2,25_P01	35.5	2.25	20	20.04	7634	4.91	0.30	11.75	59	48	40	44	1920	-	-
FROST_35,5-2,25_P02	35.5	2.25	35	34.98	8758	7.08	1.63	31.45	90	56	48	52	2688	-	-
FROST_35,5-2,25_P03	35.5	2.25	50	53.52	8712	9.28	2.48	38.44	72	48	54	51	2592	-	x
RT_H2c_P03	35.5	2.25	50*	53.52	8383	9.58	3.90	43.55	81	52	86	69	4472	-	x
				*Fixed support											

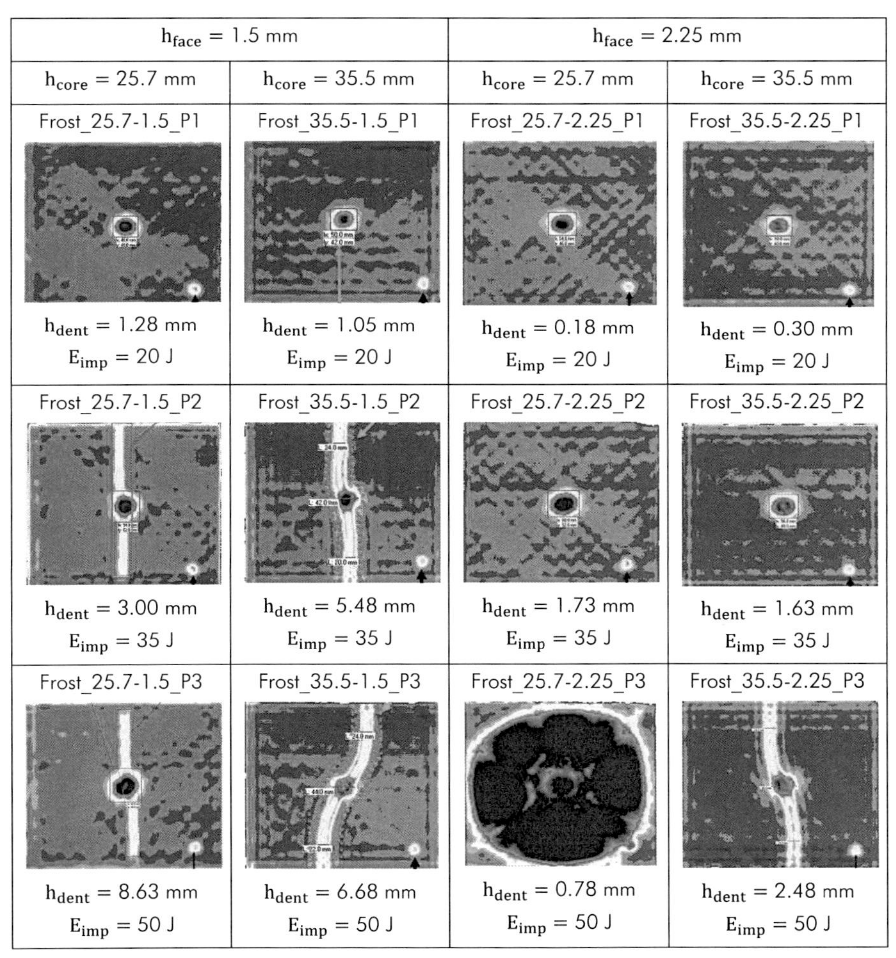

h_{face} = 1.5 mm		h_{face} = 2.25 mm	
h_{core} = 25.7 mm	h_{core} = 35.5 mm	h_{core} = 25.7 mm	h_{core} = 35.5 mm
Frost_25.7-1.5_P1	Frost_35.5-1.5_P1	Frost_25.7-2.25_P1	Frost_35.5-2.25_P1
h_{dent} = 1.28 mm E_{imp} = 20 J	h_{dent} = 1.05 mm E_{imp} = 20 J	h_{dent} = 0.18 mm E_{imp} = 20 J	h_{dent} = 0.30 mm E_{imp} = 20 J
Frost_25.7-1.5_P2	Frost_35.5-1.5_P2	Frost_25.7-2.25_P2	Frost_35.5-2.25_P2
h_{dent} = 3.00 mm E_{imp} = 35 J	h_{dent} = 5.48 mm E_{imp} = 35 J	h_{dent} = 1.73 mm E_{imp} = 35 J	h_{dent} = 1.63 mm E_{imp} = 35 J
Frost_25.7-1.5_P3	Frost_35.5-1.5_P3	Frost_25.7-2.25_P3	Frost_35.5-2.25_P3
h_{dent} = 8.63 mm E_{imp} = 50 J	h_{dent} = 6.68 mm E_{imp} = 50 J	h_{dent} = 0.78 mm E_{imp} = 50 J	h_{dent} = 2.48 mm E_{imp} = 50 J

Figure C.9 – Frost impact tests: NDI results of all specimens with clamped boundary conditions [Bie11]

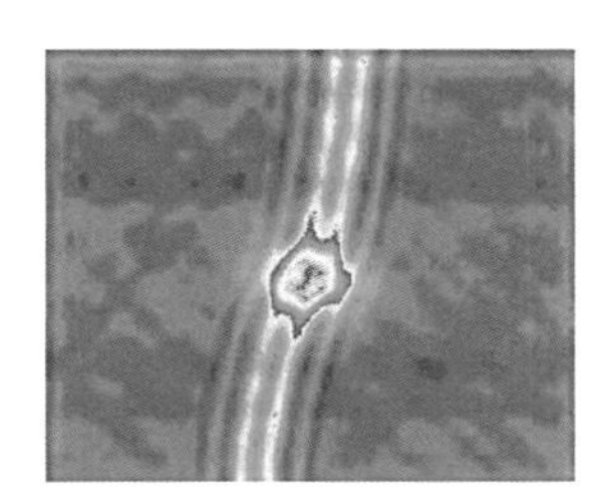

Figure C.10 – Frost impact tests: NDI results of 50 J impact test of specimen RT_H2c_P3 at -55 °C [Fri11]; the specimen was placed on a rigid support, h_{face} = 2.25 mm, h_{core} = 35.5 mm, h_{dent} = 3.9 mm

D Appendix: Simulation models

D.1 Simulation of CFRP tests

Table D.1 – modeling details of the CFRP three-point bending (3PB) tests with one element across the thickness

CFRP					
UD plies					
Material:	CFRP	Ply thickness:	0.13333 mm	Total thickness:	1.6 mm
Plies:	12	Layup:	$[(45/0/-45)_s]_2$		
Material model:	MAT54	Elastic properties:	Table 5.4	Strength properties:	Table 5.5
Element type:	Shell type 16	Hourglass type:	8		
Element edge length:		1.0 mm	Elements across laminate thickness:		1
Ply interfaces not modeled (one shell across the face sheet)					
Indenter and supports					
Material:	Steel	Density	7800 kg/m^3		
Material model:	MAT20 (rigid)	Stiffness (for contact only):	210 GPa	Poisson ratio:	0.3
Element type:	Solid type 1	Hourglass type:	-	Element edge length:	0.8 mm
Contact definitions					
Contact:	*CONTACT_AUTOMATIC_SURFACE_TO_SURFACE (indenter/support to CFRP)				
Loads and boundary conditions					
Load application:		*BOUNDARY_PRESCRIBED_MOTION_RIGID			
Indenter velocity:		1 mm/ms	Thermal load:		-150 K
Simulation time (0° / 90°)		20 / 25 ms			
Boundary conditions:		Fixed supports, no specimen clamping			

Table D.2 – modeling details of the CFRP interlaminar shear tests (ILS)

CFRP					
UD plies					
Material:	CFRP	Ply thickness:	0.132 mm	Total thickness:	1.6 mm
Plies:	12	Layup:	$[(45/0/-45)_s]_2$		
Material model:	MAT54	Elastic properties:	Table 5.4	Strength properties:	Table 5.5
Element type:	Shell type 16	Hourglass type:	8		
Element edge length:		0.5 / 1.0 mm	Elements across laminate thickness:		12
Ply interfaces					
Material:	RTM-6 resin	Thickness:	0.01333 mm		
Material model:	MAT138	Material properties:	Table B.7		
Element type:	Solid type 20	Hourglass type:	-		
Indenter and supports					
Material:	Steel	Density	7800 kg/m^3		
Material model:	MAT20 (rigid)	Stiffness (for contact only):	210 GPa	Poisson ratio:	0.3
Element type:	Solid type 1	Hourglass type:	-	Element edge length:	0.4 … 0.8 mm
Contact definitions					
Contact:	*CONTACT_AUTOMATIC_SURFACE_TO_SURFACE (indenter/supports to CFRP) *CONTACT_AUTOMATIC_SINGLE_SURFACE (CFRP to CRFP)				
Loads and boundary conditions					
Load application:		*BOUNDARY_PRESCRIBED_MOTION_RIGID			
Indenter velocity:		0.1 mm/ms	Thermal load:		-150 K
Simulation time (0° / 90°)		10 ms			
Boundary conditions:		Fixed supports, no specimen clamping			

Table D.3 – modeling details of the CFRP three-point bending tests (3PB) with twelve element across the thickness

CFRP					
UD plies					
Material:	CFRP	Ply thickness:	0.132 mm	Total thickness:	1.6 mm
Plies:	12	Layup:	$[(45/0/-45)_s]_2$		
Material model A:	MAT54	Elastic properties A:	Table B.3	Strength properties A:	Table B.4
Material model B:	MAT262	Elastic properties B:	Table B.3	Strength properties B:	Table B.5 Table B.6
Element type:	Shell type 16	Hourglass type:	8		
Element edge length:		0.5 / 2.5 mm	Elements across laminate thickness:		12
Ply interfaces					
Material:	RTM-6 resin	Thickness:	0.01333 mm		
Material model:	MAT138	Material properties:	Table B.7		
Element type:	Solid type 20	Hourglass type:	-		
Indenter and supports					
Material:	Steel	Density	7800 kg/m^3		
Material model:	MAT20 (rigid)	Stiffness (for contact only):	210 GPa	Poisson ratio:	0.3
Element type:	Solid type 1	Hourglass type:	1	Element edge length:	0.4 mm
Contact definitions					
Contact:	*CONTACT_AUTOMATIC_SURFACE_TO_SURFACE (indenter/supports to CFRP) *CONTACT_AUTOMATIC_SINGLE_SURFACE (CFRP to CRFP) *CONTACT_TIED_SHELL_EDGE_TO_SURFACE_BEAM_OFFSET (CFRP to CRFP)				
Loads and boundary conditions					
Load application:		*BOUNDARY_PRESCRIBED_MOTION_RIGID			
Indenter velocity:		1 mm/ms	Thermal load:		-150 K
Simulation time (0°-laminate / 90°-laminate)		20 / 25 ms			
Boundary conditions:		Fixed supports, no specimen clamping			

D.2 Simulation of PMI foam tests

Table D.4 – modeling details of the PMI foam indentation simulation

PMI foam					
Material:	Rohacell 71RIST				
Material model A:	MAT126	Material model B:	MAT142	Material properties:	Table B.9
Element type:	Solid type 1	Hourglass type:	9		
Element edge length:		1.0 mm			
Indenter					
Material:	Steel	Density	7800 kg/m^3	Thickness:	0.4 mm
Material model:	MAT20 (rigid)	Stiffness (for contact only):	210 GPa	Poisson ratio:	0.3
Element type:	Shell type 16	Element edge length:	0.5 mm	Hourglass type:	8
Contact definitions					
Contact:	*CONTACT_ERODING_SURFACE_TO_SURFACE (indenter to foam) *CONTACT_ERODING_SINGLE_SURFACE (foam to foam)				
Loads and boundary conditions					
Load application:		*BOUNDARY_PRESCRIBED_MOTION_RIGID			
Indenter velocity:		10 mm/ms	Thermal load:		0 K
Simulation time:		4.2 ms			
Boundary conditions:		Foam block fixed on lower side, prescribed motion on hinge arm			

D.3 Simulation of sandwich tests

Table D.5 – modeling details of the single cantilever beam (SCB) sandwich interface tests

Sandwich					
Face sheets, UD plies					
Material:	CFRP	Ply thickness:	0.13333 mm	Total thickness:	2.4 mm
Plies:	18	Layup:	$[(45/0/-45)_s]_3$		
Material model:	MAT54	Elastic properties:	Table B.3	Strength properties:	Table B.4
Element type:	Shell type 16	Hourglass type:	8		
Element edge length (x / y):		1.0 / 5.0 mm	Elements across laminate thickness:		1
Ply interfaces not modeled (one shell across the face sheet)					
PMI foam core					
Material:	Rohacell 71RIST				
Material model:	MAT142	Material properties:	Table B.9	Element edge length (z):	5.0 mm
Element type:	Solid type 1	Hourglass type:	9		
Foam core interface					
Material:	RTM-6 resin	Thickness:	0.2 mm		
Material model:	MAT138	Material properties:	Table B.13		
Element type:	Solid type 20	Hourglass type:	-		
Hinge					
Material:	Steel	Density	7800 kg/m³	Thickness:	0.4 mm
Material model:	MAT01 (elastic)	Stiffness:	210 GPa	Poisson ratio:	0.3
Element type:	Solid type 1	Hourglass type:	1	Element edge length (z):	2.5 mm
Contact definitions					
Contact:	*CONTACT_TIED_SHELL_EDGE_TO_SURFACE_BEAM_OFFSET (hinge to CFRP)				
Loads and boundary conditions					
Load application:		*BOUNDARY_PRESCRIBED_MOTION_RIGID			
Hinge velocity:		1 mm/ms	Thermal load:		-150 K
Simulation time:		26 ms			
Boundary conditions:		Fixed lower face sheet, prescribed motion on hinge			

Table D.6 – modeling details of the sandwich indentation simulation

Sandwich					
Face sheets, UD plies					
Material:	CFRP	Ply thickness:	0.132 mm	Total thickness:	1.6 mm
Plies:	12	Layup:	$[(45/0/-45)_s]_2$		
Material model A:	MAT54	Elastic properties A:	Table B.3	Strength properties A:	Table B.4
Material model B:	MAT262	Elastic properties B:	Table B.3	Strength properties B:	Table B.5 Table B.6
Element type:	Shell type 16	Hourglass type:	8		
Element edge length:		0.6 … 2.5 mm	Elements across laminate thickness:		12
Ply interfaces					
Material:	RTM-6 resin	Thickness:	0.01333 mm		
Material model:	MAT138	Material properties:	Table B.7		
Element type:	Solid type 20	Hourglass type:	-		
PMI foam core					
Material:	Rohacell 71RIST				
Material model:	MAT142	Material properties:	Table B.9	Element edge length (z):	2 mm
Element type:	Solid type 1	Hourglass type:	9		
Foam core interface					
Material:	RTM-6 resin	Thickness:	0.2 mm		
Material model:	MAT138	Material properties:	Table B.13		
Element type:	Solid type 20	Hourglass type:	-		
Indenter					
Material:	Steel	Density	7800 kg/m^3	Thickness:	0.4 mm
Material model:	MAT20 (rigid)	Stiffness (for contact only):	210 GPa	Poisson ratio:	0.3
Element type:	Shell type 16	Element edge length:	0.5 mm	Hourglass type:	8

Contact definitions	
Contact:	*CONTACT_AUTOMATIC_SURFACE_TO_SURFACE (indenter to CFRP) *CONTACT_ERODING_SURFACE_TO_SURFACE (indenter to ply interface and core) *CONTACT_AUTOMATIC_SINGLE_SURFACE (CFRP to CRFP) *CONTACT_ERODING_SURFACE_TO_SURFACE (CFRP to ply interface and core) *CONTACT_TIED_SHELL_EDGE_TO_SURFACE_BEAM_OFFSET (CFRP to CRFP and core)

Loads and boundary conditions				
Load application:	*BOUNDARY_PRESCRIBED_MOTION_RIGID			
Indenter velocity:	5 mm/ms	Thermal load:		-150 K
Simulation time:	2.5 ms			
Boundary conditions:	Fixed lower face sheet, prescribed motion on indenter			

Table D.7 – modeling details of the sandwich impact simulation

Sandwich					
Face sheets, UD plies					
Material:	CFRP	Ply thickness:	0.132 mm	Total thickness:	n*0.8 mm
Plies:	6n, n = 1…4	Layup:	$[(45/0/-45)_s]_n$		
Material model A:	MAT54	Elastic properties A:	Table B.3	Strength properties A:	Table B.4
Material model B:	MAT262	Elastic properties B:	Table B.3	Strength properties B:	Table B.5 Table B.6
Element type:	Shell type 16	Hourglass type:	8		
Element edge length:		0.6 … 2.5 mm	Elements across laminate thickness:		6*n
Ply interfaces					
Material:	RTM-6 resin	Thickness:	0.01333 mm		
Material model:	MAT138	Material properties:	Table B.7		
Element type:	Solid type 20	Hourglass type:	-		
PMI foam core					
Material:	Rohacell 71RIST				
Material model:	MAT142	Material properties:	Table B.9	Element edge length (z):	~2 mm
Element type:	Solid type 1	Hourglass type:	9		

Foam core interface					
Material:	RTM-6 resin	Thickness:	0.2 mm		
Material model:	MAT138	Material properties:	Table B.13		
Element type:	Solid type 20	Hourglass type:	-		
Impactor					
Material:	Steel	Density	7800 kg/m^3	Thickness:	0.4 mm
Material model:	MAT20 (rigid)	Stiffness (for contact only):	210 GPa	Poisson ratio:	0.3
Element type:	Shell type 16	Element edge length:	0.5 mm	Hourglass type:	8
Contact definitions					
Contact:	*CONTACT_AUTOMATIC_SURFACE_TO_SURFACE (indenter to CFRP) *CONTACT_ERODING_SURFACE_TO_SURFACE (indenter to ply interface and core) *CONTACT_AUTOMATIC_SINGLE_SURFACE (CFRP to CRFP) *CONTACT_ERODING_SURFACE_TO_SURFACE (CFRP to ply interface and core) *CONTACT_TIED_SHELL_EDGE_TO_SURFACE_BEAM_OFFSET (CFRP to CRFP and core) *CONTACT_TIED_SURFACE_TO_SURFACE (core to core)				
Loads and boundary conditions					
Load application:		*PART_INERTIA			
Impactor energy:		12 - 90 J	Thermal load:		-150 K
Impactor velocity:		2.8 - 5.8 m/s	Simulation time:		8 - 14 ms
Boundary conditions:		Picture frame fixation of upper and lower face sheets Impactor with prescribed velocity and mass			

E Appendix: Author, publications and student theses

E.1 Curriculum vitae

Personal Information:

Name:	Tim Berend Block
Date of Birth:	October 12th, 1981
Place of Birth:	Engelskirchen, Germany
Nationality:	German

School:

07/1991 – 06/2001	Gymnasium Lohmar, Gymnasium Heide-Ost and Werner-Heisenberg-Gymnasium Heide; degree: Allgemeine Hochschulreife – general qualification for university entrance
08/1998 – 06/1999	Exchange student at Cloverdale High School in California, USA

Military Service:

07/2001 – 03/2002	German Air Force, Luftwaffenausbildungsregiment 1 in Heide

University:

10/2002 – 05/2008	Technische Universität Braunschweig: Mechanical engineering (specialization in aerospace); degree: Dipl.-Ing., with distinction

Course-related activities:

08/2005 – 09/2006	Exchange student and research assistant in aeronautical and astronautical engineering at Purdue University in West Lafayette, Indiana, USA
11/2007 – 05/2008	Diploma thesis in the department structural mechanics at the German Aerospace Center (DLR), Institute of Composite Structures and Adaptive Systems, Braunschweig, Germany

Work Experience:

07/2008 – 08/2014	Faserinstitut Bremen e.V., Bremen, Germany: Researcher with focus on mechanics of composite materials and sandwich structures
09/2012 – 12/2012	Visiting researcher at Swerea SICOMP AB, Möldnal, Sweden
since 09/2014	Nordex Energy GmbH, Hamburg, Germany: Development engineer in the division blade engineering

Hamburg, October 2014

E.2 Publications of the author

This list contains only publications of the author, whose results and content overlap with this thesis and are incorporated in part into this thesis:

I. Block TB, Brauner C, Zuardy MI, Herrmann AS (2011), Advanced numerical investigation of the impact behavior of CFRP foam core sandwich structures, ECCOMAS 3rd Thematic Conference on Mechanical Response of Composites 2011, Hannover, September 2011.

II. Yokozeki T, Block TB, Herrmann AS (2011), Effects of residual thermal stresses on the energy release rates of sandwich beams for debond characterization, Journal of Reinforced Plastics and Composites 30(8), S. 699-708 (peer reviewed journal).

III. Block TB, Brauner C, Herrmann AS (2012), Impactverhalten von CFK-Schaumsandwichstrukturen (in German), VDI Wissensforum: Fachkonferenz Composites effizient Verarbeiten - Moderne Fertigungsprozesse für hohe Stückzahlen, Bremen, May 2012.

IV. Block TB, Herrmann AS (2012), Impact behavior of CFRP foam core sandwich structures, Swerea SICOMP Conference 2012, Pitea, Sweden, June 2012.

V. Block TB, Brauner C, Zuardy MI, Herrmann AS (2012), Impact damage resistance of CFRP foam core sandwich panels, 10th International Conference on Sandwich Structures (ICSS 10), Nantes, France, August 2012.

VI. Olsson R, Block TB (2013), Criteria for skin rupture and core shear cracking during impact on sandwich panels, 19th International Conference on Composite Materials (ICCM 19), Montréal, Canada, July / August 2013.

VII. Block TB, Brauner C, Herrmann AS (n.n.), Failure modes of composite foam core sandwich panels subject to low velocity impact - part I: Experimental investigation, submitted (peer reviewed journal).

VIII. Block TB, Brauner C, Herrmann AS (n.n.), Failure modes of composite foam core sandwich panels subject to low velocity impact - part II: Finite Element Analysis, submitted (peer reviewed journal).

E.3 Student theses

This thesis contains in parts results, that originated during the supervision of the following student theses:

I. Prescher J (2010), Numerische Untersuchung des Impactverhaltens von Sandwichstrukturen mit integrierten Verstärkungselementen (in German), diploma thesis, Universität Bremen.

Bisher erschienene Bände der Reihe

Science-Report aus dem Faserinstitut Bremen

ISSN 1611-3861

1	Thomas Körwien	Konfektionstechnisches Verfahren zur Herstellung von endkonturnahen textilen Vorformlingen zur Versteifung von Schalensegmenten ISBN 978-3-8325-0130-3	40.50 EUR
2	Paul Jörn	Entwicklung eines Produktionskonzeptes für rahmenförmige CFK-Strukturen im Flugzeugbau ISBN 978-3-8325-0477-9	40.50 EUR
3	Mircea Calomfirescu	Lamb Waves for Structural Health Monitoring in Viscoelastic Composite Materials ISBN 978-3-8325-1946-9	40.50 EUR
4	Mohammad Mokbul Hossain	Plasma Technology for Deposition and Surface Modification ISBN 978-3-8325-2074-8	37.00 EUR
5	Holger Purol	Entwicklung kontinuierlicher Preformverfahren zur Herstellung gekrümmter CFK-Versteifungsprofile ISBN 978-3-8325-2844-7	45.50 EUR
6	Pierre Zahlen	Beitrag zur kostengünstigen industriellen Fertigung von haupttragenden CFK-Großkomponenten der kommerziellen Luftfahrt mittels Kernverbundbauweise in Harzinfusionstechnologie ISBN 978-3-8325-3329-8	34.50 EUR
7	Konstantin Schubert	Beitrag zur Strukturzustandsüberwachung von faserverstärkten Kunststoffen mit Lamb-Wellen unter veränderlichen Umgebungsbedingungen ISBN 978-3-8325-3529-2	53.00 EUR
8	Christian Brauner	Analysis of process-induced distortions and residual stresses of composite structures ISBN 978-3-8325-3528-5	52.00 EUR
9	Tim Berend Block	Analysis of the mechanical response of impact loaded composite sandwich structures with focus on foam core shear failure ISBN 978-3-8325-3853-8	55.00 EUR

Alle erschienenen Bücher können unter der angegebenen ISBN-Nummer direkt online (http://www.logos-verlag.de) oder per Fax (030 - 42 85 10 92) beim Logos Verlag Berlin bestellt werden.